SERVICING TH

HUBBLE SPACE TELESCOPE

SPACE SHUTTLE ATLANTIS - 2009

Compiled By

DENNIS R. JENKINS and **JORGE R. FRANK**

specialtypress
PUBLISHERS AND WHOLESALERS

39966 Grand Avenue
North Branch, MN 55056 USA
(651) 277-1400 or (800) 895-4585
Fax: (651) 277-1203
www.specialtypress.com

ISBN 978-1-58007-138-3
Item Number SP138

Library of Congress Cataloging-in-Publication Data

Jenkins, Dennis R.
  Servicing the Hubble Space Telescope : Space Shuttle Atlantis / by Dennis R. Jenkins & Jorge R. Frank.
     p. cm.
  ISBN 978-1-58007-138-3
  1. Hubble Space Telescope (Spacecraft)--Maintenance and repair--History. 2. Space flights--History. 3. Atlantis (Space shuttle)--History. 4. Space flight--Planning--History. 5. Astronautics--United States--History. 6. Astronautics in astronomy. I. Frank, Jorge R. II. Title.
  QB500.268.J46 2009
  522'.2919--dc22

                                        2008036690

Printed in China
10 9 8 7 6 5 4 3 2 1

On the Front Cover: *The refurbished Hubble Space Telescope shortly after its release at the end of Servicing Mission 4. The new soft capture mechanism can be seen on the aft end of the observatory. The large aperture door at the front of the telescope is open, and the two high-gain antennas on the mid-body are fully deployed.* (NASA)

On the Back Cover (top right): *Mission specialists Michael Good (with the red-white stripe marking on the legs of his Extravehicular Mobility Unit space suit) and Mike Massimino work on Hubble during the extravehicular activity (EVA) on Flight Day 5.* (NASA)

On the Back Cover (top left): *Atlantis lifts-off from Launch Complex 39A at the Kennedy Space Center at 2:01 pm on 11 May 2009. At launch, the Space Shuttle stack weighed approximately 4.5 million pounds and generated more than 7 million pounds of thrust.* (NASA)

On the Back Cover (middle): *This NASA promotional poster was distributed throughout the NASA human spaceflight Centers and their contractors as part of the Spaceflight Awareness program that emphasizes safety.* (NASA)

On the Title Page: *Hubble secured in the payload bay of Atlantis. The remote manipulator arm (RMS) is shown deployed on the right, while the similar orbiter boom sensor system (OBSS) arm is berthed on the left of the photo.* (NASA)

Distributed in the UK and Europe by
Crécy Publishing Ltd
1a Ringway Trading Estate
Shadowmoss Road
Manchester M22 5LH England
Tel: 44 161 499 0024
Fax : 44 161 499 0298
www.crecy.co.uk
enquiries@crecy.co.uk

CONTENTS

INTRODUCTION

The story of the Hubble Space Telescope is truly an odyssey, a long journey marked by many changes of fortune. No spacecraft has survived more changes in fortune than Hubble, both over its long, troubled development period and its sometimes-troubled, sometimes-triumphant, but always interesting operational career.

An observatory situated above the distortion, clouds, and haze of the Earth's atmosphere was a centuries-long dream of astronomers. Bringing it to life was a decades-long struggle by one astronomer who was the right man in the right place at the right time.

Delayed first by its own developmental difficulties, then by the *Challenger* accident, Hubble finally took its place in low-earth orbit in 1990. The National Aeronautics and Space Administration (NASA) had touted the telescope as the most revolutionary astronomical instrument since the first built by Galileo, and expectations were high. However, the initial blurry images produced disbelief, and finally a stunned realization that something had gone terribly wrong. The subsequent investigation revealed the telescope's primary mirror had been ground incorrectly and that this error had been detected – and ignored by the contractor and the government – prior to launch. The revelation scandalized the press and the public; Hubble became an object of ridicule and a symbol for everything wrong at NASA.

Fortunately, Hubble had been designed to be periodically serviced by astronauts using the Space Shuttle. The serious issues with the telescope drove NASA to plan a first servicing mission (STS-61) far more ambitious than any Space Shuttle mission to date. The mission included an unprecedented five extravehicular activities (EVA) to replace Hubble's main camera, install corrective optics for the telescope's other instruments, and replace many other components. Not taking any chances, NASA assigned to STS-61 an all-veteran crew with more collective spaceflight experience than any previous US crew.

This caution was justified, for the Hubble debacle had caused many to question NASA's future plans. If NASA could not build a space telescope, how could they be expected to build a far larger and more complex space station? And if NASA could not pull off the five EVAs to repair Hubble, how could they pull off the dozens, possibly hundreds, of EVAs required to assemble the station in orbit? Only six months before STS-61, the space station program survived a House amendment to terminate it by a margin of 216-215. No one in Congress had to make an explicit threat in order for NASA to receive the underlying message.

As fate would have it, the launch window for STS-61 dictated that all five EVAs would occur during night in the United States. The co-compilers of this scrapbook, along with thou-

Perhaps the most famous photo taken by the Hubble Space Telescope, this false-color image of the Eagle Nebula (M16) was released in 1995. The dark pillar-like structures are columns of cool interstellar hydrogen gas and dust that are also incubators for new stars; they quickly earned the nickname, "The Pillars of Creation." The image used the Wide Field/Planetary Camera 2 (WFPC2), hence the characteristic "stair-step" shape at the upper right. The color image is constructed from three separate images taken in the light of emission from different types of atoms. Red shows emission from singly-ionized sulfur atoms. Green shows emission from hydrogen. Blue shows light emitted by doubly-ionized oxygen atoms. (NASA/STScI)

sands of others, endured five near-sleepless nights anxiously watching to see whether the telescope – and possibly the agency – would pull through. The STS-61 crew coolly rose to the occasion, restoring to Hubble its place at the pinnacle among the world's observatories, and to NASA its credibility.

Hubble continued to experience ups and downs even as it rewrote the astronomy texts. Continuing problems with the gyroscopes used to stabilize the telescope caused it to be temporarily shut down in 1999 and spurred the acceleration of the next servicing mission to restore it to operations. But the next big threat to Hubble's future would come, ironically, from the very spacecraft that serviced it and kept it alive. The 2003 *Columbia* accident resulted in a two-plus-year grounding of the surviving fleet of Orbiters, and a review of American space policy that led to a decision to dedicate remaining Space Shuttle missions to space station assembly, and retire the fleet permanently once assembly was completed. The final Hubble servicing mission, and a subsequent mission to return the telescope to Earth, were cancelled. The Space Shuttle would be replaced by an Apollo-like capsule, Orion, that would return humans to the moon but would not be capable of servicing Hubble.

As with the initial Hubble flaws, the decision to cancel the final servicing mission caused many to call NASA's other plans into question. If NASA was too risk-averse to service Hubble in low Earth orbit, what about the far higher risks of human exploration of the moon and beyond? Many within the spaceflight community felt the same way. Fortunately, in 2006, with Space Shuttle safely returned to flight and a new administrator at the helm, NASA reinstated the final servicing mission as STS-125. This recently completed mission has restored Hubble to full capability and will hopefully allow it to continue operations for many more years, even after the shuttle fleet has been retired and further servicing is no longer practical.

This history of Hubble, largely outside the scope of this short photo scrapbook, is an interesting study of the conflicts between human space flight and robotic exploration. Many, but certainly not all, science supporters point to Hubble as proof that unmanned spacecraft can produce stunning results. Nobody can seriously argue against that fact. But this ignores the immense influence of the human servicing missions. Without five teams of astronauts performing nearly unbelievable feats while orbiting the Earth at 17,500 mph, Hubble would be nothing more than a footnote of bureaucratic ineptness.

The missions also serve to underscore the unique capabilities of Space Shuttle. Capable of carrying large payloads –

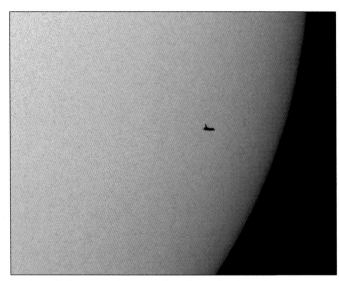

In this tightly cropped image taken from Florida, Atlantis *is seen in silhouette during solar transit on Tuesday, 12 May 2009. This image was made before* Atlantis *and the crew of STS-125 had grappled the Hubble Space Telescope.* (NASA/Thierry Legault)

Hubble, not the largest carried by the fleet, was 43.5 feet long and weighed 25,527 pounds – into orbit, and more importantly, serving as a base to service or retrieve the same payloads, these capabilities will be sorely missed as the Nation transitions back to a capsule intended to perform only a small portion of the missions originally envisioned for Space Shuttle.

This short photo scrapbook can only scratch the surface of the remarkable Hubble odyssey. To describe it completely might have challenged Homer himself. Nevertheless, others have tried; for further reading, we recommend *Hubble: Imaging Space and Time*, by David Devorkin and Robert Smith, and *The Universe in a Mirror: The Saga of the Hubble Space Telescope and the Visionaries Who Built It* by Robert Zimmerman.

We would like to thank John Grunsfeld (STS-125), Tony Landis (DFRC), Jim Ross (DFRC), Carla Thomas (DFRC), Lisa Malone (KSC), Elaine Liston (KSC), Roger Launius (NASM), Judy Rasmussen (Land's End), and Cheryl S. Gundy (STScI) for their assistance in preparing this book.

Dennis R. Jenkins
Cape Canaveral, Florida

Jorge R. Frank
Houston, Texas

THE GRANDEST OF THE GREAT OBSERVATORIES

Since the time of Galileo Galilei (1564–1642), two major issues have confounded ground-based astronomers, and both relate directly to the atmosphere that is essential to life on Earth. The first is that atmospheric conditions – such as water vapor and turbulence – disrupt the view of telescopes on the ground; this is the reason that stars seem to twinkle in the night sky. The second is that the atmosphere partially blocks or absorbs certain wavelengths of the electromagnetic spectrum, such as ultraviolet and X-rays, before they can reach Earth; this is good for plants and animals, bad for astronomers. Ideally, therefore, a telescope should be located above the atmosphere.

Hermann Oberth (1894–1989), one of the founders of modern rocketry, first proposed a space telescope in 1923. Unfortunately, it would be another six years before the German provided any significant details on the idea.

In the meantime, during 1928, Hermann Noordung, the pseudonym of Austrian Imperial Army Captain Herman Potočnik (1892–1929), published *The Problems of Space Flight* in which he showed a design for the Wohnrad space station. The primary purpose of the space station was astronomy, but only a basic description of the observatory was provided; it is assumed that Noordung was leaving the detailed definition to the astronomers.

Finally, in 1929, Oberth published *The Road to Space Travel* in which he elaborated on his earlier work with concepts for three space-based telescopes. The first was mounted on Springboard Station in a low-earth orbit. The second was Fixed-Orbit Station located in a 26,000-mile geosynchronous orbit. The third, and most ambitious, was the Asteroid-Mounted Space Telescope that used an asteroid, either left in its natural orbit or moved into an orbit around Earth, as its major component. This was intended to provide a stable platform for astronomical research, and was the first, and one of the few, suggestions to use a naturally occurring planetary body as part of a space station. Mounted on the asteroid would be a large superstructure with the objective lens of a telescope attached at the end, protected by a sunshade.

Of course, there was no technology available in 1929 to support building any of these concepts, and the idea of a space telescope would have to wait another 50 years to come to fruition.

Getting Serious

In 1946, a decade before the launch of *Sputnik*, Lyman Spitzer, Jr. (1914–1997), a professor and researcher at Yale University, wrote a paper, *Astronomical Advantages of an Extra-Terrestrial Observatory*, calling for the development of a space telescope. Spitzer was a renowned theoretical astrophysicist and believed that a large, space-based observatory could study a broad range of wavelengths not visible on Earth.

In 1965, Spitzer headed an *ad hoc* committee for the National Academy of Sciences that defined the research objectives for an orbiting telescope. The committee's 1969 report, *Scientific Uses of the Large Space Telescope*, urged the construction of an orbiting observatory that used a 120-inch (3-meter) primary mirror. Spitzer was an enthusiastic lobbyist, with both Congress and the scientific community, for the telescope. His continuing support was instrumental in convincing NASA to begin conceptual studies of the Large Space Telescope, what eventually morphed in Hubble, later in 1969.

By 1974, the conceptual design for the Large Space Telescope included a number of interchangeable instruments to study wavelengths from ultraviolet to infrared. The soon-to-be Space Shuttle would launch the telescope and service it on-orbit, ensuring a long operational life. At this point, NASA estimated it would cost $400–500 million to develop and build the telescope, not including the cost of launching or servicing it. Congress balked at the price tag and declined to fund the telescope in 1975, essentially halting the effort.

In response, Spitzer and John Bahcall, an astronomer at Princeton University, coordinated a large-scale lobbying effort by most major universities and science organizations. At the same time, they convinced NASA to invite the European Space Agency (ESA) to join the project. Perhaps most importantly, however, the scientists agreed to reduce the primary mirror from 120 inches to only 94.5 inches (2.4 meters), lowering the estimated development cost to $200 million. Satisfied with the reduction, Congress approved funding in 1977.

Even after the launch of Hubble in 1990, Spitzer remained deeply involved in the program, making important observations with the telescope while continuing to lobby for its continued operation. In addition to space astronomy, Spitzer's work greatly advanced knowledge in other fields, including stellar dynamics, plasma physics, and nuclear fusion.

When funding was approved, the Marshall Space Flight Center (MSFC) in Huntsville, Alabama, was assigned as the lead NASA center for project management, and was also responsible for leading the development of the telescope flight hardware and its support systems. The Goddard Space Flight Center (GSFC), in Greenbelt, Maryland, was assigned to lead the development of the science instruments and control the operation of the telescope.

After a competitive procurement, Marshall awarded a contract to the Perkin-Elmer Corporation (now the PerkinElmer unit of EG&G) to design and build the optical telescope assembly and fine guidance sensors for the space observatory. Perkin-Elmer was also responsible for testing the systems and delivering them to Lockheed where they would be integrated onto the spacecraft. Among the many technological advances used during construction was a computer-based laser grinding system for the primary mirror. NASA, however, worried that the new grinding system might run into trouble and contracted with Perkin-Elmer to design, fabricate, and test a smaller, 59-inch (1.5-meter) mirror to demonstrate the process worked; it did.

The demonstration notwithstanding, NASA considered the quality of the primary mirror to be a major challenge, and directed Perkin-Elmer to subcontract with the Eastman Kodak Company to fabricate a second primary mirror using more traditional methods. A team of Kodak and Itek had bid on the original mirror contract, and their proposal had called for both teammates to check each other's work, which would have almost certainly caught any polishing errors. In the end, both Kodak and Itek would build backup 94.5-inch mirrors.

The Lockheed Missiles and Space Company (now part of Lockheed Martin) was awarded a contract to manufacture the support systems module – the spacecraft – that would house the optical telescope assembly. The result of using Lockheed is that Hubble, reportedly, generally resembles a KH-11 reconnaissance satellite in overall configuration. As part of their support of the project, ESA awarded contracts in Europe to cover the development of the solar arrays that would provide power for the telescope on-orbit. By 1979, astronauts were training for the deployment mission using a full-scale mockup of the telescope in an underwater tank at Marshall.

In 1981, the Space Telescope Science Institute (STScI) was established on the Homewood campus of John Hopkins University in Baltimore, Maryland. A part of the Association of Universities for Research in Astronomy (AURA), the institute

Professor Lyman Spitzer, Jr., was one of the key figures of 20th century physics. He was instrumental in creating orbiting observatories and was one of the visionaries that supported what ultimately became the Hubble Space Telescope. (Denise Applewhite/Princeton University)

would evaluate research proposals and manage the science program. The Space Telescope European Coordinating Facility (ST-ECF), established at Garching bei München near Munich in 1984, provides similar support for European astronomers. There are approximately 100 scientists working at STScI, including 15 from ESA countries on permanent assignment. The total STScI staff consists of about 350 people.

Also during 1981, the Large Space Telescope was named after Edwin Powell Hubble (1889–1953), who made some of the most important discoveries in modern astronomy. Hubble came late to science, earning a law degree and serving in World War I before becoming an astronomer. During the 1920s, while working at the Mount Wilson Observatory in California, Hubble showed that some of the numerous distant, faint clouds of light in the Universe were actually entire galaxies. The realization that

Edwin Powell Hubble profoundly changed the understanding of the Universe by demonstrating the existence of galaxies besides the Milky Way. His observations established that the Universe continues to expand, resulting in the Big Bang theory. (National Archives)

the Milky Way was only one of many galaxies forever changed the way astronomers viewed the Universe. However, perhaps his greatest discovery came in 1929, when Hubble determined that the farther a galaxy is from Earth, the faster it appears to move away. This notion of an expanding Universe formed the basis of the Big Bang theory, which states that the Universe began with an intense burst of energy at a single moment in time and has been expanding ever since.

As with many leading-edge development efforts, NASA significantly underestimated the cost of Hubble and the engineering difficulty associated with it. By 1980, the fine guidance sensors were experiencing serious technical issues, and the severity of the challenge to keep the mirrors sufficiently free from contamination to meet the specifications in the ultraviolet spectra was just being recognized. The original $200 million budget ballooned, and ultimately Hubble cost $2.5 billion. Construction of the Perkin-Elmer mirror began in 1979 using a blank manufactured by Corning Glass Works (now Corning, Inc.) from ultra-low-expansion glass. To keep weight to a minimum, the mirror consists of 1-inch-thick top and bottom plates sandwiching a honeycomb lattice. Optical telescopes typically have mirrors polished to an accuracy of about 1/10 the wavelength of visible light, but Hubble needed to observe the shorter wavelengths in the ultraviolet spectra and was specified to be diffraction limited to take full advantage of the space environment. Therefore, the surface needed polished to an accuracy of 10 nanometers, or about 1/65 the wavelength of red light.

Mirror polishing continued until May 1981, but a report from the NASA Inspector General questioned Perkin-Elmer's managerial structure after the polishing schedule began to slip and costs increased. To save money, NASA cancelled the back-up Kodak and Itek mirrors (grinding and polishing the mirrors was completed, but they were not coated). The uncompleted Kodak mirror is now on display at the Smithsonian Institution. Interestingly, the Itek mirror was later finished and used in the 2.4-meter telescope at the Magdalena Ridge Observatory, which is used primarily to image potentially hostile spacecraft in orbit.

Ultimately, the Perkin-Elmer 94.5-inch primary mirror was completed toward the end of 1981 when the final 65-nanometer-thick reflective coating of aluminum and 25-nanometer-thick protective coating of magnesium fluoride were applied. However, the entire optical telescope assembly was not put together until 1984. The original suite of science instruments was delivered in 1983, and the entire telescope was completed in 1985. These delays caused the planned December 1983 launch date to slip to October 1986. But the unthinkable was about to happen.

The Great Observatories

Astronomers need to analyze radiation throughout the electro magnetic spectrum. This presents a problem since the Earth's atmosphere partially or completely blocks many of these spectra. To overcome this limitation of ground-based instruments, NASA developed four space-based Great Observatories designed to study different wavelengths (visible, gamma rays, X-rays, and infrared) of radiation. An important aspect of the Great Observatories was overlapping the operational phases of the missions to enable astronomers to make contemporaneous observations of an object at different spectral wavelengths.

The first Great Observatory – and the best known to the public – is the Hubble Space Telescope. Its domain extends from the ultraviolet, through the visible, and into the near-infrared. In technical terms, these wavelengths extend from 1,200 ångström in the ultraviolet (1 ångström is 1 hundred-millionth of a centimeter) to 24,000 ångströms in the near-infrared.

Compared to ground-based telescopes, Hubble is not particularly large. With a 94.5-inch (2.4-meter) primary mirror, Hubble would at most be considered a medium-size telescope on Earth where more than 40 telescopes have larger mirrors. For instance, the W. M. Keck Observatory on the summit of the Mauna Kea volcano in Hawaii has a pair of 400-inch (10-meter) telescopes equipped with adaptive optics.

The newest, large ground-based telescopes use computer-controlled adaptive optics that can compensate for much of the moisture and turbulence in the atmosphere. These instruments rival or exceed Hubble's vision in some areas. However, even the best adaptive optics will not allow a ground-based telescope to observe wavelengths that have been filtered out by the atmosphere, making Hubble, in its 380-mile (612-kilometer) orbit above Earth, a unique instrument.

The second Great Observatory was the Compton Gamma Ray Observatory, designed to collect data on some of the most violent physical processes in the Universe. The observatory was named after Arthur Holly Compton (1892–1962), a Nobel laureate for his work with gamma ray physics. Compton was built by TRW (now Northrop Grumman Space Technology) in Redondo Beach, California. Following 14 years of development and testing, the observatory was launched by STS-37 using the Space Shuttle *Atlantis* on 5 April 1991. At the time of its launch, the 35,000-pound (15,876-kilogram) observatory was the heaviest astrophysical payload ever flown. Compton was deployed in a 280-mile (450-kilometer) low-earth orbit to avoid the Van Allen radiation belt. The observatory operated until it was safely deorbited on 4 June 2000.

The third Great Observatory was the Chandra X-Ray Observatory, used to observe black holes, quasars, and high-temperature gases throughout the X-ray portion of the electromagnetic spectrum. Chandra was launched into a 170-mile (275-kilometer) orbit by STS-93 using the Space Shuttle *Columbia* on 23 July 1999. After deployment, an Inertial Upper Stage placed the telescope into a 6,000 x 87,000-mile (9,650 x 140,000-kilometer) elliptical orbit. The observatory was named in honor of Indian-American physicist Subrahmanyan Chandrasekhar (1910–1995), a Nobel laureate best known for determining the mass limit for white dwarfs to become neutron stars. The observatory was originally known as the Advanced X-Ray Astrophysics Facility (AXAF) and was assembled by TRW (now Northrop Grumman Space Technology) in Redondo Beach, California. Chandra uses the most precisely shaped and smoothest mirrors ever constructed; if the surface of Earth were as smooth as the mirrors, the highest mountain would be 78 inches (2 meters) tall. Chandra images are twenty-five times sharper than the best previous X-ray telescope.

The Spitzer Space Telescope was the fourth and final Great Observatory. Originally known as the Space Infrared Telescope Facility (SIRTF), the observatory was orbited by a Delta II 7920EH expendable launch vehicle on 25 August 2003. Unlike the other Great Observatories, which are in earth orbits, Spitzer was in a heliocentric orbit around the sun with a 1-year period. Consisting of a 33-inch (0.85-meter) telescope and three cryogenically cooled science instruments, Spitzer was the largest infrared telescope launched into space. The observatory operated at wavelengths between 3 and 180 microns (1 micron is one-millionth of a meter) that are normally blocked by the atmosphere and cannot be observed from the Earth. Unlike most telescopes, which are named by a board of scientists, SIRTF was named after Lyman Spitzer, Jr., by a public contest, much to the delight of science educators. The planned mission duration was 2.5 years, but ultimately the onboard liquid helium supply, needed to keep the infrared detectors at their optimum temperature, lasted until 15 May 2009, almost 6 years.

THE HUBBLE SPACE TELESCOPE

Hubble is a two-mirror reflecting telescope similar to most Earth-based telescopes built in the last 100 years. These two-mirror instruments are generally called Cassegrain telescopes, after the French cleric who first postulated the design. However, Hubble is a special type of Cassegrain telescope, called a Ritchey-Chretien variant, that has better optical performance over a larger image plane. The mirrors in the Ritchey-Chretien design are slightly more aspheric (have a greater departure from a pure spherical shape) than in the standard Cassegrain type. For Hubble, the primary mirror is a 94.5-inch (2.4-meter) concave hyperboloid and the secondary mirror is a 12-inch (0.3-meter) convex hyperboloid. This makes the Hubble a little less than half the size of the 60-year-old Hale Telescope at the Palomar Observatory outside San Diego, California, and a quarter the size of several newer telescopes.

The optical telescope assembly consists of the primary and secondary mirrors, support trusses, and the focal plane structure. Light enters the aperture and travels down the main baffle (a surface that eliminates stray, unwanted, light). Light is reflected by the primary mirror; because of its concave shape, the primary mirror converges the light to a secondary mirror through a secondary baffle. The secondary mirror reflects the still-converging light back to the primary mirror through a central baffle. The light travels through a hole in the center

Hubble's 94.5-inch primary mirror prior to being installed in the optical telescope assembly. Light reflects off the main mirror then bounces off a smaller secondary mirror through the hole (shown with a cover) in the center of the primary mirror. (NASA/STScI)

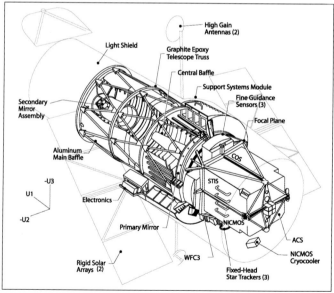

The major components of the Hubble Space Telescope are shown here. Note how the science instruments sit in a group behind the primary mirror. This is the final configuration of the spacecraft, shown with the SM4 instruments installed. (NASA/STScI)

of the primary mirror to reach the focal plane and the science instruments.

The optical telescope assembly is 17.5 feet (5.3 meters) long and 9.6 feet (2.9 meters) in diameter. The 252-pound (114-kilogram) truss was manufactured from graphite-epoxy, the lightest and strongest material available at the time. In addition, graphite-epoxy resists expanding or contracting during the 100 degF (38 degC) changes the spacecraft experiences as it moves from Sun exposure to shadow on each orbit.

The spacecraft in which the optical telescope assembly is housed was a major engineering achievement. It has to withstand frequent passages from direct sunlight into the darkness of Earth's shadow, resulting in major temperature changes, while remaining stable enough to allow extremely accurate pointing of the telescope. Originally, a shroud of multi-layer insulation on the outside of the spacecraft was used to keep the temperature within the telescope stable. However, during the servicing missions, stainless steel sheets were installed over part of Hubble's exterior to provide additional thermal protection to some equipment bays, replacing the multi-layer insulation that had slowly degraded over time with exposure to the harsh space environment. Overall, the telescope is 43.5 feet (13.2 meters) long, 14 feet (4.2 meters) in diameter, and weighs 25,527 pounds (11,574 kilograms).

The pointing control system (PCS) aligns Hubble so that the telescope remains locked on a target that might be thousands of light-years away. The PCS is designed for pointing to within 0.01 arcsecond (a 60th part of a minute of arc) and is capable of holding a target for up to 24 hours while Hubble continues to orbit the Earth at 17,500 mph, although other constraints prohibit this from actually happening. If the telescope were in Los Angeles, it could hold a laser pointer on a dime in San Francisco.

It would have been impractical to use conventional rocket thrusters to maneuver Hubble since that would have required frequent visits by the Space Shuttle to deliver propellant. Instead, Hubble exploits some very basic physics to maneuver and look at different parts of the sky. The telescope uses six gyroscopic rate-sensing units (which, like a compass, always point in the same direction) and four free-spinning reaction wheel assemblies.

Hubble was designed to use three of the six gyroscopes to meet its very precise pointing requirements, with the other three held as spares. Gyros have limited lifetimes, and

Astronauts train on a full-scale underwater mockup of the telescope in the 40-foot-deep Neutral Bouyancy Laboratory (NBL) at the Johnson Space Center (JSC). This mockup has been used for all six Space Shuttle missions to Hubble. (National Archives)

Joseph Tanner (on the robotic arm) and Gregory Harbaugh replace a fine guidance sensor during Servicing Mission 2 (STS-82) in 1997. Each fine guidance sensor is 5 feet wide, 3 feet long, and weighs 485 pounds. (National Archives)

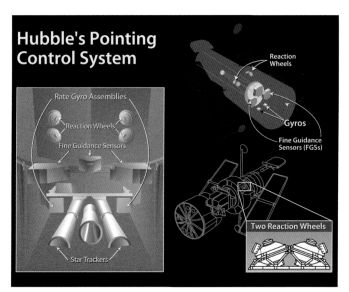

Hubble's Pointing Control System

Rate Gyro Assemblies

Reaction Wheels

Fine Guidance Sensors

Star Trackers

Reaction Wheels

Gyros

Fine Guidance Sensors (FGSs)

Two Reaction Wheels

As could be expected, the Hubble pointing control system is a major contributor to its success. The reaction wheels provide the momentum to move the telescope, while the star trackers and fine guidance sensors lock-onto stars apparently near the desired target. The gyros detect and correct for very fine motion, such as jitter. (NASA)

following a thorough analysis and testing, engineers determined the telescope could operate productively on two gyros. After the implementation of three new control modes in the main computer, and major changes to Hubble's planning and scheduling system at the STScI, two-gyro operations began in 2005. By operating on two gyros, with the other gyros turned off until needed, Hubble will be able to operate longer after the final servicing mission, which installed six new gyros.

In addition, a magnetic sensing system measures Hubble's position relative to Earth's magnetic field and a pair of coarse sun sensors measures Hubble's orientation to the Sun and assists in deciding when to open and close the aperture door.

The four reaction wheels are Hubble's "steering" system. According to Newton's Third Law of Motion, every action has an equal but opposite reaction. Therefore, as Hubble accelerates its reaction wheels in one direction, the telescope rotates in the opposite direction. Flight software commands the reaction wheels to spin, accelerating or decelerating up to 3,000 rpm, as needed to point the telescope toward a new target. Four magnetic torquers are used to manage reaction wheel speed. Reacting against Earth's magnetic field, the torquers reduce the reaction wheel speed, thus managing angular

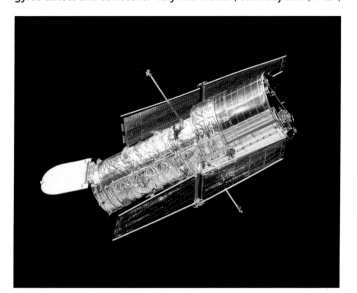

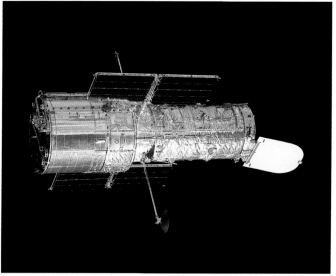

Two views of Hubble showing the different solar arrays. The photo at left was taken during SM2 in 1997 and shows the 40-foot-long SA2 solar arrays that were installed during SM1 in 1993. Although these resolved much of the jitter associated with the original solar arrays, they still introduced some vibration as the telescope transited from cold darkness into warm daylight. The photo at right shows the 25-foot-long SA3 solar arrays installed during SM3B. Although one-third smaller, they produce 30 percent more power, are less susceptible to extreme temperatures, and their smaller size reduced the effects of atmospheric drag on the spacecraft. (NASA)

momentum. Since the rotation axes of the four reaction wheels point in different directions, controllers are able to use different combinations of them to point the telescope toward any location in the sky.

Three fine guidance sensors are Hubble's targeting devices. They aim the telescope by locking onto "guide stars" and measuring the position of the telescope relative to the target. The sensors provide the precise reference point from which the telescope can begin repositioning.

All of Hubble's power comes from a pair of 25-foot-long (7.6-meter), 8-foot-wide (2.4-meter) solar arrays. Each array provides 2,800 watts of electricity. Some of the energy generated is stored in onboard batteries so the telescope can operate while it's in Earth's shadow (which is about 36 minutes out of each 97-minute orbit). Fully charged, each battery can sustain the telescope in normal science operations mode for 7.5 hours, or five orbits. The original six batteries lasted from the 1990 deployment to the final servicing mission in 2009. The crew of STS-125 replaced both battery modules (each containing three batteries); combined with the power system enhancements made during SM3B this will provide ample power margins for the remainder of Hubble's lifetime.

Hubble is operated by the Flight Operations Team in the Space Telescope Operations Control Center (STOCC) at the Goddard Space Flight Center, Maryland. Science operations are controlled by the Space Telescope Science Institute (STScI) on the Homewood campus of John Hopkins University in Baltimore, Maryland. (NASA)

Hubble is lifted into the upright position at Kennedy Space Center in preparation for its 1990 launch aboard the Space Shuttle Discovery. A close look at this image reveals a portion of the 225 linear feet of handrails installed around the outside of the spacecraft for astronauts to grip during servicing mission spacewalks. (NASA)

SCIENCE INSTRUMENTS

The aft section of Hubble can accommodate five science instruments and three fine guidance sensors. Science instruments are classified according to how they are mounted. Axial instruments are about the size and shape of a telephone booth and are mounted in the rear of the telescope along the optical axis where light from the mirrors enters the instruments directly. There are four axial bays, each enclosed by a set of double doors that allow on-orbit access to service or replace the instrument. Radial instruments are mounted in a ring around the front of the axial instruments and receive light via "pickoff mirrors." Fine guidance sensors that control the pointing of the telescope occupy three of the radial bays. The sensors are in the focal plane structure, at right angles to the optical path and 90 degrees apart. When two fine guidance sensors lock-on to guide stars to provide pointing information for the telescope, the third sensor serves as a science instrument to measure the position of stars in relation to other stars. A science instrument uses the remaining radial bay.

The complement of science instruments has changed on almost every servicing mission. As initially deployed in 1990 by STS-31R, the instruments included:

Fine Guidance Sensors. Located in the radial bay. These sensors lock-on to guide stars that help the telescope obtain the exceedingly accurate pointing necessary for observation of astronomical targets. These instruments can also be used to obtain highly accurate measurements of stellar positions.

Wide Field/Planetary Camera 1 (called WF/PC initially, but later WFPC1). Located in the radial bay. Pronounced "wiffpick," this imaging camera was used to obtain high-resolution images of astronomical objects over a relatively wide field of view and a broad range of wavelengths (1,150 to 11,000 ångströms). The instrument was built by the Jet Propulsion Laboratory (JPL) and incorporated a set of 48 filters isolating spectral lines of particular astrophysical interest. The

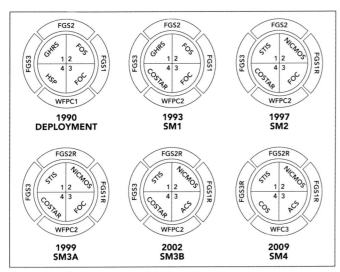

Representative configuration of the Hubble science instruments after each servicing mission. The radial bays are around the outside, while the axial bays are in the center, numbered from top left to bottom left in this drawing. (Dennis R. Jenkins)

An astronaut removes the Goddard High Resolution Spectrograph during SM2 in 1997 so that the Space Telescope Imaging Spectrograph could be installed. All of the science instruments in the axial bays are generally similar in size and shape. (STScI/NASA)

camera operated in either wide field mode that captured the largest area, or planetary mode that captured a smaller area but with higher resolution. The instrument's functionality was severely impaired by the defects of the main mirror, but nevertheless, it produced uniquely valuable high-resolution images of relatively bright astronomical objects, allowing for a number of discoveries to be made by Hubble even in its aberrated condition.

Goddard High Resolution Spectrograph (GHRS). Located in axial bay 1. Manufactured at Goddard, this instrument obtained high-resolution spectra of bright ultraviolet targets from 1,150 to 3,200 ångström.

Faint Object Spectrometer (FOS). Located in axial bay 2. Manufactured by Martin Marietta, this instrument made spectroscopic observations of very faint sources from the near ultraviolet to the near-infrared (1,150 to 8,000 ångströms). This instrument also had a polarimeter for the study of the polarized light from these sources.

Faint Object Camera (FOC). Located in axial bay 3. Built by Dornier GmbH and supplied by the European Space Agency, this instrument imaged a very small field of view and very faint targets from the near-ultraviolet to the near-infrared (1,150 to 6,500 ångströms).

High Speed Photometer (HSP). Located in axial bay 4. Built at the University of Wisconsin at Madison, this instrument measured very fast brightness changes in diverse objects, such as pulsars, from the near-ultraviolet to the visible. The HSP could take up to 100,000 measurements per second with a photometric accuracy of about 2 percent or better. The design was novel in that, despite being able to view through a variety of filters and apertures, it had no moving parts. Unfortunately, the HSP could not be used successfully due to the optical problems with the telescope.

The first servicing mission in December 1993 installed the corrective optics required to compensate for the defective primary mirror, and the crew of STS-61 also replaced two science instruments.

Wide Field/Planetary Camera 2 (WFPC2). Located in the radial bay. Built by the Jet Propulsion Laboratory. Since this instrument was located in the radial bay, it could not use the COSTAR corrective optics and included its own built-in corrective optics. Like the original WF/PC, the WFPC2 was used to obtain high-resolution images of astronomical objects over a relatively wide field of view and a broad range of wavelengths (1,150

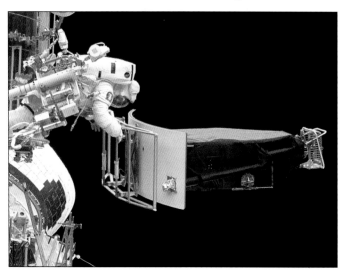

Astronauts remove the Wide Field and Planetary Camera (WFPC1) to replace it with its more powerful WFPC2 successor during SM1 in 1993. The camera, shaped something like a grand piano, weighed 610 pounds and was located in the radial bay. (STScI/NASA)

The Advanced Camera for Surveys (ACS) being processed inside the clean room at the Vertical Processing Facility (VPF) at the Kennedy Space Center. None of the science instruments are particularly interesting-looking from the outside. (NASA)

to 11,000 ångströms). The primary difference was that the new instrument used greatly improved sensors. The images from the WFPC2 have a curious stair-step shape since the instrument consisted of four cameras, each recording a separate image that represented one part of the overall view. One of the cameras recorded a magnified view of the section it observed, allowing finer detail to be recorded for that section. During image processing the magnified view was reduced to the pro-

portion of the other three, resulting in one small image and three larger images. The stair-step shape emerged when the four images were stitched together. This instrument replaced the first-generation WF/PC.

Corrective Optics Space Telescope Axial Replacement (COSTAR). Located in axial bay 4. COSTAR was not a science instrument, per se, but instead consisted of mirrors that refocused the aberrated light from the optical system for the GHRS, FOC, and FOS instruments in the remaining axial bays (but could not correct the light for the radial bay instruments). It was built by Ball Aerospace and replaced the High Speed Photometer, which was returned to Earth.

Servicing Mission 2, STS-82 in February 1997, replaced two of the three remaining first-generation instruments, leaving only the Faint Object Camera from the initial configuration.

Space Telescope Imaging Spectrograph (STIS). Located in axial bay 1. Built by Ball Aerospace, this instrument has the ability to simultaneously obtain high-resolution spectra from many different points along a target. This instrument operated at ultraviolet and visible wavelengths (1,150 to 10,300 ångströms) and replaced the first-generation GHRS.

Near-Infrared Camera/Multi-Object Spectrometer (NICMOS). Located in axial bay 2. Built by Ball Aerospace, this instrument provides imaging capabilities in the 0.8–2.5 micron wavelength range. NICMOS has three adjacent but not contiguous cameras, designed to operate independently, each with a dedicated sensor array at a different magnification scale. To be sensitive in the near-infrared, the sensors must be cooled to a very low temperature. Unfortunately, issues with the solid-nitrogen cooling system forced controllers to turn this instrument off in January 1999 until the crew of SM3B could install an improved cooling system in March 2002. NICMOS replaced the first-generation FOS.

The third servicing mission, SM3A in December 1999, did not replace or service any of the science instruments, concentrating instead on the condition of the spacecraft.

The fourth servicing mission, SM3B in March 2002, installed a new cooling system for the Near-Infrared Camera/Multi-Object Spectrometer and the crew of STS-109 also replaced one science instrument.

Various parts of the SM3B payload sit in the Vertical Processing Facility. At the rear are the folded-up Solar Array 3 panels in thier carrier, with the Advanced Camera for Surveys in the middle. At the front is the cradle Hubble sits on while being serviced. (NASA)

Advanced Camera for Surveys (ACS). Located in axial bay 3. Built in cooperation between Johns Hopkins University, Ball Aerospace, Goddard, and the STScI, this instrument replaced the first-generation FOC and also largely superseded the capabilities of the WFPC2, except for certain narrow-band filters. The ACS operates in wavelengths from the far ultraviolet to visible light (1,150 to 11,600 ångströms), making it capable of studying some of the earliest activity in the Universe and is the most popular instrument for observers. ACS contains a trio of imaging instruments: the wide field camera, the high-resolution camera, and the solar blind camera. With a field of view twice that of WFPC2, the wide field camera conducts broad surveys of the Universe to study the nature and distribution of galaxies for clues about how the Universe evolved. The high-resolution camera takes extremely detailed pictures of the inner regions of galaxies, searching neighboring stars for possible planets. It has also taken close-up images of the planets in our own solar system. The solar blind camera, which blocks visible light to enhance ultraviolet sensitivity, focuses on hot stars or planets radiating ultraviolet wavelengths. An electrical short in 2007 disabled the wide field camera and high-resolution camera, leaving only the solar blind camera function.

The fifth, and final, servicing mission in May 2009 left behind the most-capable instruments yet installed on the telescope. The crew of STS-125 repaired the Space Telescope Imaging Spectrograph and the Advanced Camera for Surveys, and installed two new science instruments:

Wide Field Camera 3 (WFC3). Located in the radial bay. Built by Goddard and Ball Aerospace, this instrument uses many parts recovered from the original WF/PC and spares manufactured for WFPC2. The instrument incorporates the most popular filters from the ACS and earlier WFPC instruments and is designed to be a versatile camera capable of imaging astronomical targets over a very wide wavelength range with a large field of view. The detailed "planetary" mode has been deleted, hence the subtle change in nomenclature.

Cosmic Origins Spectrograph (COS). Located in axial bay 4. Built at the Astrophysics Research Lab in the Center for Astrophysics and Space Astronomy at the University of Colorado at Boulder, this ultraviolet spec-

trograph is optimized for observing faint point sources with moderate spectral resolution. This instrument is designed for high throughput, medium-resolution spectroscopy of point sources, allowing the efficient observation of numerous faint extragalactic and galactic ultraviolet (1,150 to 3,000 ångströms) targets. It replaced the second-generation corrective optics (COSTAR) that has not been used since the removal of the first-generation GHRS, FOC, and FOS instruments.

The Wide Field Camera 3 (WFC3), is prepared for a dress rehearsal of EVA activities by a technician in a large cleanroom at the Goddard Space Flight Center. This view gives a good sense of the size and shape of the radial bay instruments. (NASA)

HUBBLE OPERATIONS

The Flight Operations Team in the Space Telescope Operations Control Center (STOCC) operates Hubble by sending commands via the Tracking and Data Relay Satellite System (TDRSS) to the telescope's onboard computers. One of the main computers handles commands that point the telescope and other system-wide functions; the other computer controls the science instruments and sends their data to the ground through TDRSS. The majority of these operations are programmed in advance, but controllers can interact with the spacecraft in real time as needed.

Hubble is situated in a 350-mile (563-kilometer), low-Earth orbit so that it can be reached by the Space Shuttle for servicing missions, but this means that the Earth occults most astronomical targets for slightly less than half of each orbit. Observations also cannot take place when the telescope passes through the South Atlantic Anomaly due to elevated radiation levels, and there are sizable observation exclusion zones around the Sun, Earth, and Moon. The solar avoidance angle is about 50 degrees to keep sunlight from illuminating any part of the optical telescope assembly while Earth and Moon avoidance keeps bright light out of the fine guidance sensors.

The Flight Operations Team is busier than usual during servicing missions. Shortly after the Space Shuttle is launched, controllers command the telescope to stop normal science operations, close the aperture door, and stow the high-gain antennas. Changes to the telescope are tested as the servicing mission progresses, mostly to identify any issues the crew needs to correct before returning to Earth. At the end of each servicing mission, the Flight Operations Team deploys the high-gain antennas and opens the aperture door. They then reactivate the telescope and conduct the servicing mission orbital verification test before beginning science operations.

Technically, anybody can apply to use Hubble; there are no restrictions on nationality or academic affiliation. However, in reality, there are so many proposals from recognized astronomers that anybody without serious credentials does not stand a chance. Each year astronomers from dozens of countries vie for precious minutes of Hubble's time and submit proposals to the STScI where experts from the astronomical community review them. Calls for proposals are issued roughly annually, with each cycle lasting approximately one year. More than 1,000 proposals are reviewed and approximately 200 are selected, representing roughly 20,000 individual observations.

Astronomers may also submit "target of opportunity" proposals, where observations are scheduled if a transient event covered by the proposal occurs during the scheduling cycle. In addition, approximately 10 percent of the telescope schedule is allocated as director's discretionary time. This time is not included in the normal call for proposals, and is typically awarded to study unexpected transient phenomena such as supernovae.

In 1986, the first director of STScI, Riccardo Giacconi, announced that he intended to devote some of his discretionary time to amateur astronomers. The total time was only a few hours per cycle, but excited great interest among the amateur community. A committee of leading amateur astronomers reviewed proposals for observation time, and only proposals that had genuine scientific merit were selected. In total, 13 amateur astronomers were awarded time on the telescope, with observations being carried out between 1990 and 1997. Unfortunately, subsequent funding reductions at STScI eliminated this worthwhile program.

The STScI converts information, such as which instrument to use, what filter to use, and how long the exposure should be, into a detailed list of second-by-second instructions that is loaded onto the Hubble computers a few days before the scheduled observation. As Hubble completes a particular observation, it converts the starlight into digital signals that are transmitted through TDRSS to a ground station at White Sands, New Mexico, via landlines to Goddard, and then to the Space Telescope Science Institute for processing. Astronomers can download archived data via the internet and analyze it from anywhere in the world.

Hubble is orbiting the Earth at 17,500 mph (28,200 kph) and Earth is moving around the Sun at 67,000 mph (107,800 kph). The telescope completes an orbit around the Earth every 97 minutes. As might be expected, keeping the telescope aimed at an object several light-years away is a challenge. To accomplish this, a pair of "guide stars," whose apparent positions are near the science target, are selected from the *Guide Star Catalog*, which lists the brightness and positions of 200 million stars. Guide star information is combined with the detailed observation instructions and transmitted to Hubble's onboard computer via TDRSS. The computers then send the data to the fine guidance sensors, which lock-on to the guide stars and maintain the precise pointing needed for an observation.

Photo Techniques

Hubble is noted for its beautiful and often bizarre color images. The color, however, is not necessarily representative of how the object would appear to the human eye.

A typical Hubble image is made from a combination of black-and-white digital images representing different parts of the electromagnetic spectrum. To record what an object looks like at a certain wavelength, Hubble uses special filters that allow only a certain range of wavelengths through to the imaging sensors. Since the sensors can detect light outside the visible light spectrum, the use of filters allows scientists to study features only visible in ultraviolet or infrared wavelengths. Finished color images are combinations of black-and-white exposures to which color has been added during image processing. These colors are assigned to enhance an object's detail or to visualize what ordinarily could never be seen by the human eye, and are not intended to produce a "real" representation of the object.

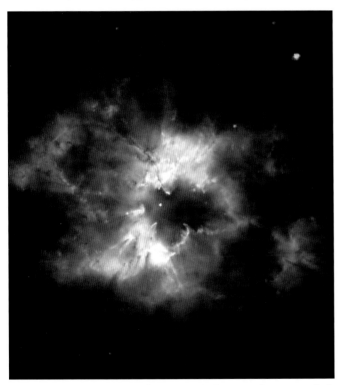

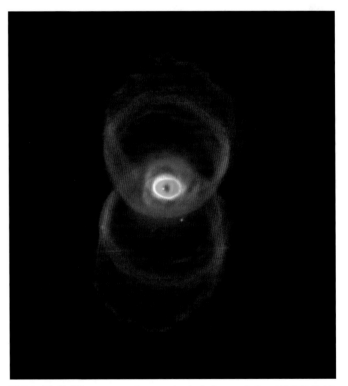

This October 1999 image of NGC 2440 shows a planetary nebula ejected by a dying star. The central star of NGC 2440 is one of the hottest known, with a surface temperature near 360,000 degF. The complex structure of the nebula suggests to some astronomers that there have been periodic oppositely directed outflows. The nebula is also rich in dust clouds, some of which form long, dark streaks pointing away from the central star. NGC 2440 lies about 4,000 light-years from Earth in the direction of the constellation Puppis. (STScI)

This is MyCn18, a young planetary nebula located about 8,000 light-years from Earth. Taken with the WFPC2 on 16 January 1996, this picture has been composed from three separate images in the light of ionized nitrogen (represented by red), hydrogen (green), and double-ionized oxygen (blue). According to one theory for the formation of planetary nebulae, the hourglass shape is produced by the expansion of a fast stellar wind within a slowly expanding cloud that is more dense near its equator than near its poles. (STScI)

In the most active starburst region in the local Universe lies a cluster of brilliant, massive stars, known as Hodge 301. Seen in the lower-right-hand corner of this image, Hodge 301 is inside the Tarantula Nebula in the Large Magellanic Cloud. Many of the stars in Hodge 301 are so old that they have exploded as supernovae, spewing material out into the surrounding region at velocities approaching 200 miles per second. (STScI)

This composite image of the barred spiral galaxy NGC 1512 was taken in all wavelengths from ultraviolet to infrared using the FOC, WFPC2, and NICMOS instruments. NGC 1512 is in the southern constellation of Horologium, some 30 million light-years from Earth. The galaxy spans 70,000 light-years and has a 2,400-light-year-wide circle of infant star clusters, called a "circumnuclear" starburst ring, at its core. (STScI)

It is impossible to study the Milky Way galaxy from a distance, so astronomers study similar galaxies instead. Like the Milky Way, NGC 3949 has a blue disk of young stars peppered with bright pink star-birth regions. In contrast to the blue disk, the bright central bulge is made up of mostly older, redder stars. NGC 3949 lies about 50 million light-years from Earth and is one of the larger galaxies in a loose cluster of some six or seven dozen galaxies located in the direction of the Big Dipper, in the constellation Ursa Major (the Great Bear). This image was created in October 2001 using separate WFPC2 exposures through blue, visible, and near-infrared filters. (STScI)

Thousands of sparkling young stars are nestled within the giant nebula NGC 3603, a prominent star-forming region in the Carina spiral arm of the Milky Way, about 20,000 light-years from Earth. The image reveals stages in the life cycle of stars. Powerful ultraviolet radiation and fast winds from the bluest and hottest stars have blown a big bubble around the cluster. Moving into the surrounding nebula, this torrent of radiation sculpted the tall, dark stalks of dense gas, which are embedded in the walls of the nebula. These gaseous monoliths are a few light-years tall and point to the central cluster. The stalks may be incubators for new stars. The image spans roughly 17 light-years and was taken on 29 December 2005 by the ACS. (NASA/ESA/STScI/AURA)

Spiral galaxy Messier 74 (M74), also called NGC 628, is a stunning example of a "grand-design" spiral galaxy that is viewed by Earth observers nearly face-on. Its perfectly symmetrical spiral arms emanate from the central nucleus and are dotted with clusters of young blue stars and glowing pink regions of ionized hydrogen (hydrogen atoms that have lost their electrons). These regions of star formation show an excess of light at ultraviolet wavelengths. Tracing along the spiral arms are winding dust lanes that also begin very near the galaxy's nucleus and follow along the length of the spiral arms. M74 is located roughly 32 million light-years from Earth in the direction of the constellation Pisces, the Fish. In its entirety, it is estimated that M74 is home to about 100 billion stars, making it slightly smaller than the Milky Way. (NASA/ESA/STScI)

This 2001 image of an unusual edge-on galaxy, ESO 510-G13, reveals details of its warped dusty disk and shows how colliding galaxies spawn the formation of new generations of stars. The dust and spiral arms of normal spiral galaxies, like the Milky Way, appear flat when viewed edge-on. By contrast, ESO 510-G13 has an unusual twisted disk structure, first seen in ground-based photographs obtained at the European Southern Observatory (ESO) in Chile. The galaxy lies in the southern constellation Hydra, roughly 150 million light-years from Earth. (STScI)

To the surprise of astronomers, this galaxy, NGC 4622, appears to be rotating in the opposite direction to what they expected. NGC 4622 has an outer pair of winding arms full of new stars (shown in blue). Astronomers are puzzled by the clockwise rotation because of the direction the outer spiral arms are pointing. Most spiral galaxies have arms of gas and stars that trail behind as they turn, but this galaxy has two "leading" outer arms that point toward the direction of clockwise rotation. To add to the conundrum, NGC 4622 also has a "trailing" inner arm that is wrapped around the galaxy in the direction opposite its rotation. Astronomers suspect that NGC 4622 interacted with another galaxy since its two outer arms are lopsided, suggesting that something disturbed it. NGC 4622 resides 111 million light-years from Earth in the constellation Centaurus. The pictures were taken in May 2001 with the WFPC2 using ultraviolet, infrared, blue, and green filters. (STScI)

Hubble can also image planets in our solar system, and took this picture of Mars on 26 June 2001, when the planet was approximately 43 million miles from Earth. A large dust storm is shown around the northern polar cap (top of image). (STScI)

The same area of Mars in December 2007 when the planet was 55 million miles from Earth. The planet appears free of any dust storms during this closest approach; however, there are significant clouds visible in both the northern and southern polar cap regions. (STScI)

This unusual true-color image of Jupiter was taken using WFPC2 on 17 February 2007 in support of the unmanned New Horizons planetary mission. Jupiter's trademark belts and zones of high- and low-pressure regions appear in crisp detail. (STScI/AURA)

Three black shadows cast by three of Jupiter's moons – Io (the white dot), Callisto (blue), and Ganymede. Io's shadow is just above center and to the left; Ganymede's is on the left edge; and Callisto's near the right edge. Callisto is out of the image to the right. (STScI)

The moons Mimas, Enceladus, and Dione are highlighted against Saturn's disk. Dione is at left, Mimas in the middle, and Enceladus is on the right. The Image reveals the planet's rings tilted nearly edge-on toward the Sun, an event that occurs once every 15 years. Because of this alignment, Dione and Enceladus are casting long shadows on the rings. The image was taken on 17 November 1995 with the WFPC2 using blue, green, and red filters. (STScI/NASA/ESA/E. Karkoschka at the University of Arizona)

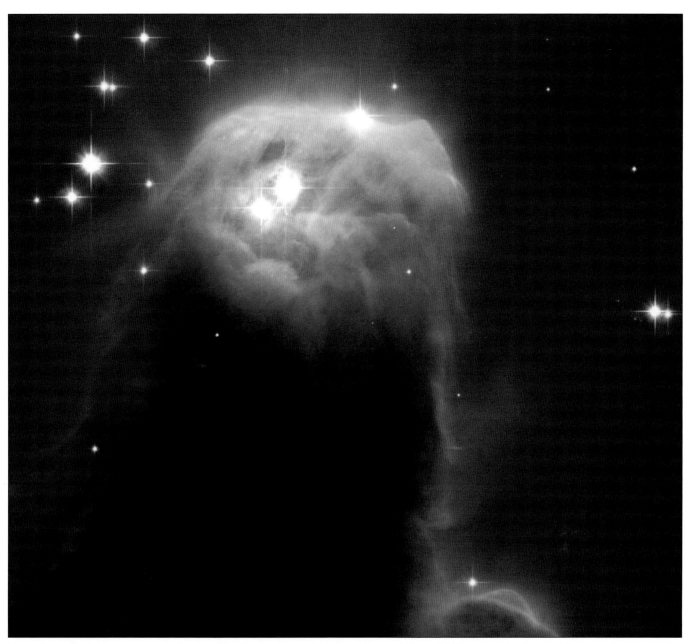

When it released this photo on 30 April 2002, the STScI stated, "Resembling a nightmarish beast rearing its head from a crimson sea, this monstrous object is actually an innocuous pillar of gas and dust." It was a good description. Called the Cone Nebula (NGC 2264) – so named because, in ground-based images, it has a conical shape – this giant pillar resides 2,500 light-years from Earth in the constellation Monoceros. This picture was taken by the ACS and shows the upper 2.5 light-years of the nebula; the entire nebula is 7 light-years tall. Astronomers believe that these pillars are incubators for developing stars and are common in large regions of star birth. The photo was constructed from three separate images taken with blue, near-infrared, and hydrogen-alpha filters. (STScI)

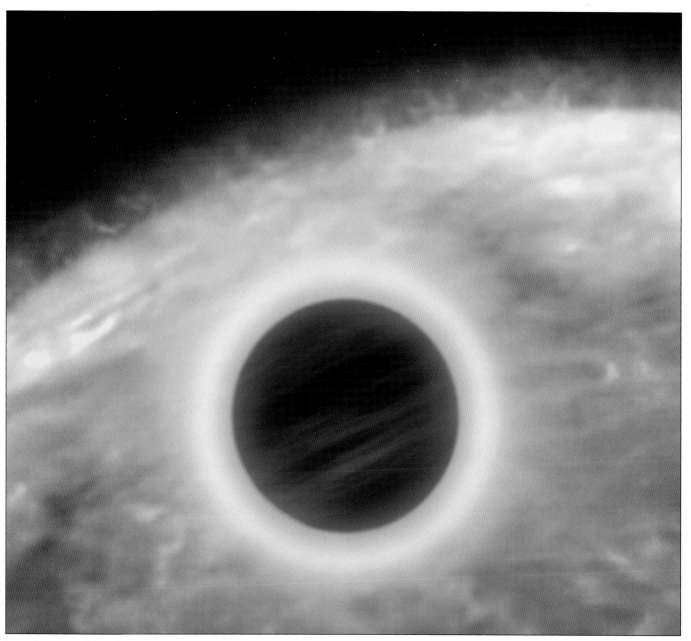

In early 2007 Hubble allowed astronomers to study for the first time the layer-cake structure of the atmosphere of a planet, designated HD 209458b, orbiting another star. The planet orbits so close to its star and gets so hot that its gas is streaming into space, making the planet appear to have a comet-like tail. Researchers determined that the layer in the planet's upper atmosphere "is actually a transition zone where the temperature skyrockets from about 1,340 degF to 25,540 degF, which is hotter than the Sun," said Gilda Ballester of the University of Arizona in Tucson. The gas escapes the planet's gravitational pull at a rate of 10,000 tons per second, more than three times the rate of water flowing over Niagara Falls, but the planet will not disappear for at least another 5 billion years. (Artist concept by NASA/STScI)

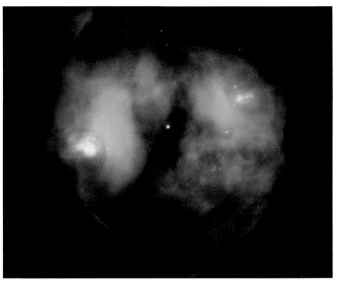

A glowing bubble of gas and dust encircle NGC 2371, a dying star 4,300 light-years from Earth in the constellation Gemini. The remnant star visible at the center is the core of the former red giant with a surface temperature of 240,000 degF. (STScI/AURA)

This close-up view shows only a 3-light-year-wide portion of the entire Carina Nebula, which has a diameter of over 200 light-years. Located 8,000 light-years from Earth, the nebula can be seen in the southern sky with the naked eye. This image shows a region in the Carina Nebula between two large clusters of some of the most massive and hottest known stars. (STScI)

This violent and chaotic mass of gas and dust is the remnant of nearby supernova N63A. The remnant is a member of N63, a star-forming region in the Large Magellanic Cloud (LMC). Visible from the southern hemisphere, the LMC is an irregular galaxy lying 160,000 light-years from Earth. The image is a color representation of data taken in 1997 and 2000 with the WFPC2. (STScI)

This pair of galaxies (NGC 3808 on the right and NGC 3808A), known collectively as Arp 87, was first cataloged by astronomer Halton Arp in the mid-1960s using the 200-inch Palomar Observatory telescope. NGC 3808 is a nearly face-on spiral galaxy with a bright ring of star formation and several prominent dust arms. Stars, gas, and dust flow from NGC 3808 to form an enveloping arm around NGC 3808A, a spiral galaxy seen edge-on surrounded by a rotating ring that contains stars and interstellar gas clouds. (NASA/ESA/STScI)

Sirius A is the brightest star in our nighttime sky, and is only 8.6 light-years from Earth. Sirius B (the small dot at lower left) , a white dwarf, is very faint because of its tiny size, only 7,500 miles in diameter. The cross-shaped diffraction spikes and concentric rings around Sirius A, and the small ring around Sirius B, are artifacts produced within Hubble's imaging system. (STScI)

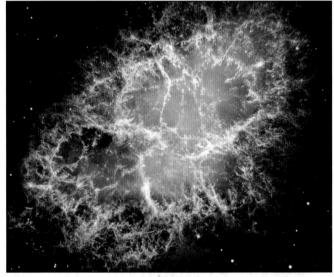

This is a mosaic image of the Crab Nebula, a 6-light-year-wide expanding remnant of a star's supernova explosion 6,500 light-years from Earth. Japanese and Chinese astronomers witnessed this violent event nearly 1,000 years ago in 1054. The orange filaments are the hydrogen remains of the star and the rapidly spinning neutron star embedded in the center of the nebula emits the bluish glow. (STScI)

DEPLOYMENT – STS-31R

As Lockheed and Perkin-Elmer worked through the technical issues associated with developing Hubble, the planned 1983 launch date came and went with little fanfare. The telescope was finally manifested for launch in October 1986. However, on the cold morning of 28 January 1986, *Challenger* was destroyed just over a minute into its flight. The Space Shuttle fleet was grounded for 32 months while the Rogers Commission investigated the accident and the program made changes to correct serious defects in hardware and procedure.

The finished Hubble parts were moved into storage, and workers continued to tweak the telescope, improving the batteries and upgrading other systems. In particular, the ground and onboard computer systems were revamped using better technology that became available through the electronics revolution that was taking place during the early 1980s. There is little doubt that the telescope that was ultimately launched was vastly superior to one that could have been launched in 1983; but one major problem had been completely overlooked.

After a 42-month delay, the Hubble Space Telescope was finally launched on STS-31R using Space Shuttle *Discovery* on 24 April 1990. (There had been an STS-31 – called 61-B by public affairs – prior to the *Challenger* accident, so this mission was designated STS-31R, for "reflight.") The crew included Loren J. Shriver, commander; Charles F. Bolden, Jr., pilot; Steven A.

Hawley, mission specialist (MS) 1; Bruce McCandless II, MS2; and Kathryn D. Sullivan, MS3.

The telescope weighed 25,517 pounds (11,574 kilograms) and carried five instruments: the Wide Field/Planetary Camera, Goddard High Resolution Spectrograph, Faint Object Camera, Faint Object Spectrometer, and High Speed Photometer. Hubble was released into a 380-mile (612-kilometer) orbit inclined 28.45-degrees on 25 April. This was the highest altitude flown by Shuttle to date. Two IMAX cameras recorded the telescope's deployment to film *Destiny in Space*, which was released in 1994. *Discovery* landed on the concrete runway at Edwards AFB, California, on 29 April 1990.

Almost immediately, it became apparent that something was wrong. The original specification called for the telescope to focus 90 percent of the light it captured into a circle 0.1 arc-second in radius; on-orbit tests showed the light was spread out into a circle that measured 0.7 arcseconds. While the images produced by Hubble were clearer than those of ground-based telescopes, they weren't anywhere near as pristine as promised. They were blurry.

By 25 June, engineers and scientists had discovered the problem. Hubble's primary mirror, polished over the course of a full year, had a flaw called "spherical aberration" that caused the light bouncing off the center of the mirror to focus in a different place than light bouncing off the edge. The state-of-the-art computer-controlled laser grinding system used by Perkin-Elmer had ground the outer edge of the mirror too flat by 2.2 microns. The tiny flaw, only 1/50 the thickness of a sheet of paper, was enough to distort the view. While scientists could still study the cosmos and make significant discoveries, the distortion meant that much of the original mission could not be fulfilled.

A commission headed by Lew Allen, director of the Jet Propulsion Laboratory, investigated the root cause of the problem. Released in November 1990, their report found that the device used to measure the exact shape of the mirror had been assembled incorrectly. While the mirror was being polished, Perkin-Elmer had analyzed its surface with two other devices that (correctly) indicated that the mirror was suffering from spherical aberration, but the company ignored these test results. While the Allen Commission heavily criticized Perkin-Elmer, it also criticized NASA for a lack of oversight and for relying totally on test results from a single instrument.

During a discussion forum in April 2008, former NASA administrators Robert A. Frosch (1977–1981) and James M. Beggs (1981–1985) both indicated that not conducting independent testing on the primary mirror was the worst decision made during their respective tenures. Frosch also commented that he was not comfortable with the concept of "co-prime contractors" where Lockheed built the spacecraft and Perkin-Elmer built the optics. Regardless of 20-20-hindsight, Hubble was crippled. While Kodak and Itek had each manufactured back-up mirrors for Hubble, it would be impossible to replace the mirror on orbit, and too expensive and time-consuming to bring the telescope temporarily back to Earth for a refit.

Fortunately, engineers were dealing with a well-understood optical problem – albeit in a wholly unique situation. And they had a solution. A series of small mirrors could be used to intercept the light reflecting off the primary mirror, correct for the flaw, and bounce it to the telescope's science instruments. Because of the way the instruments were mounted, the correction would only work for the axial instruments, not the radial bays. Fortunately, a new Wide Field/Planetary Camera 2 already included relay mirrors that could completely cancel the aberration of the primary mirror. Since this was the only radial science instrument, it was not an issue. The existing axial instruments would use the Corrective Optics Space Telescope Axial Replacement (COSTAR) installed in place of the High Speed Photometer on the first servicing mission. In essence, Hubble would receive a very expensive pair of glasses.

Discovery *climbs toward space carrying the Hubble Space Telescope on 24 April 1990 from the Kennedy Space Center. The entire vehicle weighed 4,514,665 pounds at launch, including the 25,517-pound telescope and its support equipment.* (NASA)

Photographed at the moment of deployment, the remote manipulator on Discovery releases Hubble into its 380-mile orbit on 25 April 1990. The photo was taken by the IMAX Payload Bay Camera, mounted on the port side of Discovery in Bay 12. (NASA)

SERVICING MISSION 1 (SM1) – STS-61

Hubble had been designed so that it could be regularly serviced by Space Shuttle, but the problems with the mirror meant the first servicing mission assumed a much greater importance, as astronauts would have to carry out extensive work on the telescope to install the corrective optics.

Although engineers and scientists understood the mirror problem, it still took time to build the corrective optics and to schedule another Space Shuttle mission to install them. It would be more than three years, until 2 December 1993, before *Endeavour* carried seven astronauts into orbit for a mission to fix Hubble. The crew included: Richard O. Covey, commander; Kenneth D. Bowersox, pilot; Kathryn C. Thornton, MS1; Claude Nicollier, MS2; Jeffrey A. Hoffman, MS3; F. Story Musgrave, MS4/payload commander; and Thomas D. Akers, MS5. *Endeavour* carried 17,662 pounds (8,011 kilograms) of new equipment and supplies to Hubble.

The mission lasted almost 11 days, and the crew made five extravehicular activities (EVA). To be on the safe side, the flight plan allowed for two additional EVAs, but these were ultimately unnecessary. To accomplish this without too much fatigue, the EVAs were shared between two alternating pairs of astronauts, a technique used on most future Hubble servicing missions.

The primary task was to replace the High Speed Photometer with the Corrective Optics Space Telescope Axial

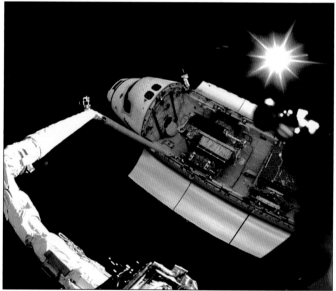

Standing on the end of the remote manipulator arm, one of the astronauts photographed Endeavour *backdropped against the blackness of space, with the Sun displaying a rayed effect. The Orbiter still had an internal airlock, and the door to the payload bay is open. (NASA)*

Replacement (COSTAR) system to correct the spherical aberration of the main mirror. COSTAR was a telephone booth-sized instrument designed and built by Ball Aerospace that placed five pairs of corrective mirrors, some only 1 inch in diameter, in front of the Faint Object Camera, Faint Object Spectrometer, and Goddard High Resolution Spectrometer. The mission also replaced the original Wide Field/Planetary Camera with the newer Wide Field/Planetary Camera 2, which had built-in corrective optics and significantly improved ultraviolet performance.

In addition, the SM1 crew replaced the solar arrays, solar array drive electronics, magnetometers, coprocessors for the flight computer, two rate sensor units, and two gyroscope electronic control units. Of these, the new solar arrays (called SA2) were the most welcome because they reduced a "jitter" caused by excessive solar panel flexing during the telescope's orbital transition from cold darkness into warm daylight.

Finally, the telescope's orbital altitude was boosted to compensate for the orbital decay from 3 years of drag in the upper atmosphere. On 9 December, the crew redeployed the telescope and *Endeavour* landed at the Kennedy Space Center on 13 December 1993.

This successful mission not only improved Hubble's vision – which led to a string of remarkable discoveries in a very short time – but it also validated the effectiveness of on-orbit servicing using Space Shuttle. The first new images from Hubble's fixed optics were released on 13 January 1994. The pictures were beautiful; their resolution, excellent. Hubble was transformed into the telescope that had been originally intended.

High over Madagascar, Hubble is berthed in Endeavour's *payload bay. The crew used TV cameras to survey the spacecraft before sending two pairs of astronauts on five separate spacewalks to perform a variety of servicing tasks.* (NASA)

During EVA-2, one of the original solar arrays failed to retract and Kathy Thornton jettisoned the array by hand while in a foot restraint on the end of the RMS. The other solar array was successfully retracted, stowed, and returned to Earth. (NASA)

SERVICING MISSION 2 (SM2) – STS-82

STS-82 was the first "routine" servicing mission for Hubble, and in fact, the only one of the five servicing missions that could be called that. No major crisis needed tended to, but a variety of new equipment would be installed to enhance the capabilities of the telescope. *Discovery* was launched on the night of 11 February 1997 carrying a crew of seven: Kenneth D. Bowersox, commander; Scott J. Horowitz, pilot; Joseph R. Tanner, MS1; Steven A. Hawley, MS2; Gregory J. Harbaugh, MS3; Mark C. Lee, MS4/payload commander; and Steven L. Smith, MS5.

The primary task was to replace the Goddard High Resolution Spectrometer and Faint Object Spectrometer with the Space Telescope Imaging Spectrograph and Near-Infrared Camera/Multi-Object Spectrometer. Both instruments had optics that corrected for the flawed primary mirror. In addition, they featured technology that wasn't available when scientists designed the original instruments in the late 1970s, opening a broader viewing window for Hubble.

The Space Telescope Imaging Spectrograph searched for black holes by studying the gas dynamics around galactic centers, measured the distribution of matter in the Universe by studying quasar absorption lines, and used its high sensitivity and spatial resolution to study star formation in distant galaxies and perform spectroscopic mapping of solar system objects. The Near-Infrared Camera/Multi-Object Spectrometer

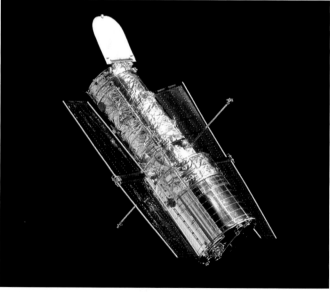

Looking as good as new, Hubble is redeployed following the STS-82 servicing mission. The main optical aperture is at the top of the photo, with the protective door in the open position. The poles protruding from the middle carry the high-gain antennas. (NASA)

consisted of three cameras that used a 230-pound block of nitrogen ice to cool their infrared detectors. The instrument was capable of infrared imaging and spectroscopic infrared observations of astronomical objects. Unfortunately, shortly after it was installed, an unexpected thermal expansion resulted in part of the nitrogen ice block coming into contact with an optical baffle. This led to an increased warming rate for the instrument and reduced its original expected lifetime of 4.5 years to about 2 years.

In addition to installing the new instruments, STS-82 installed a refurbished fine guidance sensor, replaced a reel-to-reel engineering science tape recorder with a 25-pound solid state recorder that could store ten times as much data (12 gigabytes instead of 1.2 gigabytes), and replaced one of four reaction wheel assemblies with a refurbished spare. Astronauts also repaired a data interface unit and replaced the second solar array drive electronics unit (one was replaced during SM1; the unit that was returned from orbit was refurbished and used on SM2 to replace the second unit). As had SM1, this servicing mission boosted Hubble's orbit to compensate for atmospheric drag. *Discovery* returned to the Kennedy Space Center on 21 February 1997.

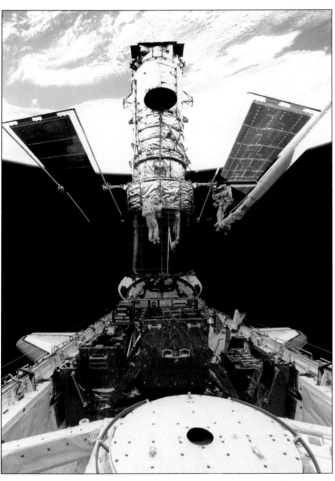

Backdropped against Australia, Steven Smith (center) and Mark Lee (on the arm) conducted a survey of the hand rails and insulation on Hubble during the fifth spacewalk. The external airlock in the forward payload bay is visible in the foreground. (NASA)

Lee (with red stripe on his suit leg) and Smith working in the payload bay. Note the fold-out checklist attached to Lee's arm. Gold foil is used extensively as thermal insulation to mitigate the wide swings in temperature as the vehicle goes from sunlight to darkness. (NASA)

SERVICING MISSION 3 (SM3A) – STS-103

Planning for the third Hubble servicing mission, to be conducted in June 2000, began almost as soon as STS-82 landed. However, what was originally conceived as a routine servicing mission turned more urgent in February 1999 when the third of six rate-sensing units (gyroscopes) failed on the telescope; the first gyro had failed shortly after SM2 in 1997 and the second failed in mid-1998. Three working gyros, each spinning at a constant 19,200 rpm, were required to aim the telescope precisely enough to conduct science operations. NASA decided to split the third servicing mission into two parts, SM3A and SM3B, and to launch the first as soon as possible to replace all six gyros with improved units.

A "call-up" mission – one that had not been manifested in the normal, time-consuming manner – was quickly approved. This was one of the few times the Space Shuttle Program has used the procedure.

On 13 November 1999, the situation became more critical when a fourth gyro failed and the Flight Operations Team at Goddard began putting Hubble into a dormant state called safe mode. Essentially, Hubble went to sleep while it waited for help. Controllers closed the aperture door to protect the optics and aligned the spacecraft to ensure that Hubble's solar panels would receive adequate power from the Sun. Ironically, the same day the fourth gyro failed, *Discovery* was making her slow trip

Hubble sits securely on the flight support structure (FSS) in the pay-load bay after it was berthed at 19:46:15 UTC on 21 December 1999 during SM3A. Note the discoloration on the exterior surface of the telescope, and how dull the solar arrays are. (NASA)

from the Vehicle Assembly Building (VAB) to the launch pad in preparation for the launch of SM3A.

The seven crewmembers for STS-103 included: Curtis L. Brown, commander; Scott J. Kelly, pilot; Steven L. Smith, MS1/payload commander; Jean-Francois Clervoy, MS2; John M. Grunsfeld, MS3; C. Michael Foale, MS4; and Claude Nicollier, MS5. Unfortunately, a series of scrubs delayed *Discovery's* launch until 19 December 1999; the mission would be only the second time that a U.S. crew had been in space for Christmas.

In addition to replacing all six gyros, the crew replaced a fine guidance sensor and one computer during three 6-hour spacewalks. The 478-pound fine guidance sensor was a refurbished unit that had been returned by SM2. The new 70-pound computer (a 25 MHz radiation-hardened Intel 486 with 2 megabytes of RAM) was 20 times faster and had six times the memory of the original unit. In addition, a voltage improvement kit was installed to protect the batteries from overcharging whenever the telescope entered safe mode. The repair mission also installed a new S-band transmitter for communicating with TDRSS. Hubble has two identical transmitters and can operate with only one; the new transmitter replaced one that failed in 1998. During SM2, one of the reel-to-reel data recorders had been replaced with a solid-state recorder; SM3A replaced one more.

During SM2, astronauts noted that the detected damage to some of the telescope thermal insulation blankets that protect the telescope from the harsh space environment. Years of exposure had taken a toll on Hubble protective multi-layer insulation, and some areas were torn or broken. The crew of SM3A installed new outer blanket layer (NOBL) covers over three areas of insulation to protect it. These covers prevent further degradation of the insulation and maintain normal operating temperatures.

The *Discovery* crew deployed Hubble back into orbit on Christmas day and landed at the Kennedy Space Center on 27 December 1999.

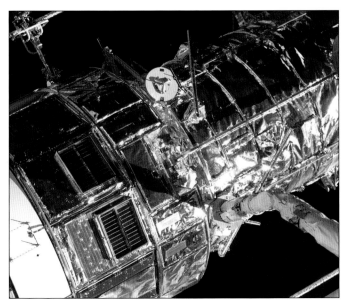

Hubble has two grappling fixtures on the exterior surface that the Orbiter's remote manipulator arm can grab. Here, the arm has grabbed a fixture toward the bottom of the photo; another fixture may be seen toward the top center of the image. (NASA)

Claude Nicollier (left) and Michael Foale replace a fine guidance sensor with a refurbished unit that has an enhanced on-orbit alignment capability. A 70mm camera inside Discovery's cabin was used to record this image. Note the Moon at lower right. (NASA)

SERVICING MISSION 3, REDUX (SM3B) – STS-109

The decision to split the third servicing mission into two parts meant that SM3A had been a remedial mission – it restored Hubble to working condition. The second part of the mission, designated SM3B, would install the new Advanced Camera for Surveys that was ten times as sensitive as the Faint Object Camera it replaced.

The original 19 November 2001 launch date for STS-109 was pushed back to give engineers time to evaluate a reaction wheel assembly that had been acting up on Hubble, as well as by several scrubs for weather and mechanical issues. *Columbia* was finally launched on 1 March 2002 on an 11-day mission that would include five spacewalks. The seven crewmembers included: Scott D. Altman, commander; Duane G. Carey, pilot; John M. Grunsfeld, MS1/payload commander; Nancy J. Currie, MS2; Richard M. Linnehan, MS3; James H. Newman, MS4; and Michael J. Massimino, MS5.

The SM1 crew had replaced the original solar arrays with improved SA2 units that had powered the telescope for over 8 years. On SM3B, these were replaced by a new generation of solar arrays (SA3) that were based on units developed for the Iridium communications satellites. Although one-third smaller, they produce 30 percent more power, are less susceptible to extreme temperatures, and their smaller size reduced the effects of atmospheric drag on the spacecraft.

The mechanism on the right is how Hubble is secured to its frame in the payload bay of the Orbiter during servicing missions. The hooks at the bottom grab the rod above them as the telescope is lowered into position. There are three of these capture points. (NASA)

An additional benefit is that the additional power allows all of the science instruments to be run simultaneously and reduced a vibration problem that occurred when the old, less rigid arrays entered and left direct sunlight.

The crew also replaced the power control unit that distributes electricity from the solar arrays and batteries to other parts of the telescope. Replacing the original 11-year-old power control unit required the Flight Operations Team at Goddard to completely power-down Hubble for the first time since it was launched in 1990. The crew also installed a new outer blanket layer (NOBL) over a fourth area of degraded thermal insulation; SM3A had installed three other blankets.

The Near-Infrared Camera/Multi-Object Spectrometer installed by SM2 in 1997 had a finite operational life because it used a 230-pound block of nitrogen ice to cool its infrared detectors; this ice was depleted about 23 months after it was installed and the instrument was turned off. SM3B installed an experimental cryogenic cooling system that uses ultra-high-speed microturbines, the fastest of which spins at over 200,000 rpm, to cool the detectors to –315 degF (–193 degC). Hubble's engineering team successfully demonstrated this technology in 1998 aboard STS-95. Although not as cold as

the original nitrogen ice block, the temperature was more stable and not time limited. Astronauts also replaced one of the four reaction wheel assemblies. *Columbia* landed at the Kennedy Space Center on 12 March 2002.

After the return of STS-109, NASA began planning the fourth servicing mission (SM4), with a targeted launch in mid-2004. Everything changed on 1 February 2003 when *Columbia* disintegrated during entry from the STS-107 mission.

With his feet secured on a platform connected to the remote manipulator arm, Mike Massimino works with James Newman during the second spacewalk. Inside Columbia, *Nancy Currie controlled the arm while the others replaced a reaction wheel assembly.* (NASA)

Newman, floating in the payload bay of Columbia, *shows his spacesuit with solid red stripes around the legs; Massimino wore alternating red and white stripes (see photo at left). This allowed the astronauts to be easily identified.* (NASA)

CANCELLATION AND REINSTATEMENT

When *Columbia* (OV-102) lifted-off from the Kennedy Space Center as STS-107 on 16 January 2003, the fifth Hubble servicing mission (SM4) was manifested on OV-102 in early 2005. *Columbia* was the logical vehicle to use for the servicing mission since it was somewhat heavier than the other Orbiters and therefore could not carry as much payload to the International Space Station. But fate intervened. A piece of insulating foam came off the External Tank during the launch of STS-107, and impacted the leading edge of the Orbiter's left wing. Unknown to the flight crew and ground controllers, it was a fatal condition; *Columbia* disintegrated during entry on 1 February 2003, killing the crew of seven.

The Space Shuttle fleet was grounded pending the report of the Columbia Accident Investigation Board (CAIB), headed by retired four-star admiral Harold W. Gehman, Jr. Like all missions, SM4 was postponed. When the CAIB finally released their report in August 2003, NASA interpreted various recommendations as implying that all future Space Shuttle missions would fly to the International Space Station. Although the CAIB did not specifically restrict other missions, the Board recommended inspection and repair techniques be developed that did not appear to be possible in the near-term. In consultation with his senior advisors, NASA Administrator Sean O'Keefe cancelled the final Hubble servicing mission.

Oddly, there was not, initially, a formal announcement of the cancellation, although a little-noticed article in *The Washington Post* on 14 January 2004 mentioned the decision. On 16 January, O'Keefe visited the Goddard Space Flight Center and told the assembled Hubble staff he had cancelled the mission because of safety concerns. Word immediately leaked to the news media and, that evening, Dr. John M. Grunsfeld, NASA's chief scientist and an astronaut who had flown on two previous Hubble servicing missions, answered questions from the press. Grunsfeld stated, "To respond responsibly to the Columbia Accident Investigation Board's recommendations that we have inspection and repair techniques would have involved the development of a tremendous amount of technology that looks like it would have been too big of a challenge."

The major concerns was that the crew of a Hubble mission could not seek "safe haven" aboard the International Space Station in case of major problems with the Orbiter that might prevent a safe entry. Contrary to popular science fiction, changing the orbital inclination of a spacecraft requires a tremendous amount of propellant, and the ISS and Hubble are in vastly different orbits. Without safe haven or certified techniques for fixing thermal protection system damage in orbit, O'Keefe determined that a Hubble flight was too dangerous.

One of the primary critics of the cancellation was Senator Barbara A. Mikulski (D-MD), a supporter of all things NASA, and Goddard projects in particular. To defray some of the criticism, O'Keefe asked ADM Gehman, the chairman of the CAIB, to "review the matter and offer his unique perspective." Gehman responded on 5 March 2004, emphasizing that the views expressed in the letter were his own, and not the collective thoughts of the Board (which had been disbanded after the final report was issued). Uncharacteristically for a man used to making life-or-death decisions, Gehman largely avoided the issue, concluding "I suggest only a deep and rich study of the entire gain/risk equation can answer the question of whether an extension of the life of the wonderful Hubble telescope is worth the risks involved."

In February 2005, Sean O'Keefe resigned from NASA to become Chancellor of Louisiana State University, and Dr. Michael D. Griffin succeeded him on 13 April 2005.

In response to O'Keefe's original decision, the White House had removed funding for the Hubble mission from the FY06 budget. However, In December 2005, after the National Academy of Sciences issued a report calling on NASA to reinstate the mission, Congress appropriated $291 million for some type of Hubble Servicing mission.

Unexpectedly, given the usually blasé attitude of the public toward space, O'Keefe's cancellation had touched off a storm of protest from the general public, as well as scientists, prompting NASA to look into using the appropriated funds for a robotic servicing mission. The goals of the unmanned mission included attaching a propulsion module that could drive Hubble to a safe, targeted entry at the end of its useful life. But such a robotic mission ultimately was deemed too technically challenging and, at something over $1 billion, too expensive. Griffin cancelled work on the robotic mission on 29 April 2006, but appeared to favor a manned mission if the risks could be made acceptable.

"For any given mission, I would say that the Hubble servicing represents the highest priority utilization of a single Shuttle mission that I can conceive," Griffin said in a 2005 interview. "Because servicing the Hubble is something only the Shuttle can do, it's only one flight and is, therefore, I think a very high agency priority if we can do it technically."

Griffin's decision would come following the third return-to-flight mission after the *Columbia* accident. The general feeling within NASA was that although foam still came off the External Tank, sufficient progress had been made toward eliminating it that the risks had been reduced to an acceptable level. In addition, the orbiter boom sensor system (OBSS) and its sensors had proven capable of inspecting the underside of the Orbiter and wing leading edge for damage. There was still no viable repair capability for the reinforced carbon-carbon (RCC) leading edge that had failed on *Columbia*, but several techniques were available for repairing the tiles on the bottom of the Orbiter.

On 31 October 2006, Griffin reinstated the servicing mission, deciding the scientific value of the orbiting observatory justified the additional cost – and higher risk – of the flight. Appearing at Goddard before an audience of Hubble workers, Griffin said, "We are going to add a servicing mission to the Hubble Space Telescope to the Shuttle manifest and fly it before it retires." Mikulski, a staunch supporter of Hubble and the long-awaited rescue mission, rose to her feet and led a standing ovation. The reborn servicing mission was manifested as STS-125 on *Discovery* for May 2008. As a condition of approving the mission, Griffin required a dedicated launch-on-need (LON) rescue flight be available if needed.

At the end of 2006, NASA was rearranging the flight manifest to make the best use of the available Orbiters and processing facilities. As part of a general realignment first shown on the 8 January 2007 manifest, the STS-125 Hubble mission was switched from *Discovery* to *Atlantis* and slipped until 11 September 2008. At the time, this was expected to be the last flight of *Atlantis* before it was retired, but subsequent manifests have shown additional flights for the Orbiter.

The changes made to the External Tanks to minimize foam shedding during launch and ascent proved more difficult than expected to manufacture. Because the Michoud Assembly Facility (MAF) was unable to produce tanks as quickly as expected, it became necessary for NASA to further delay STS-125. Although the tank (ET-127) for the Hubble mission was ready in time for the September launch, the tank (ET-129) needed for the contingency LON flight of STS-400 was not, prompting a delay to 8 October. Further complicating preparations, tropical storm Fay pounded the central Florida coast for most of the week of 18 August, forcing NASA to close the Kennedy Space Center for several days.

At the end of September 2007, Warner Brothers Pictures and IMAX Corporation announced that an IMAX 3D camera would travel with STS-125 for a new film that will chronicle the story of the Hubble telescope. IMAX has made a number of movies centered on space, including *Destiny in Space*, *The Dream is Alive*, *Mission to Mir*, *Blue Planet*, *Magnificent Desolation*, and *Space Station 3D*.

Happier days. NASA Administrator Sean O'Keefe posing with Columbia *before the STS-109 (SM3B) mission.* Columbia *would be lost on its next mission in early 2003.* (NASA)

SERVICING MISSION 4 (SM4) – STS-125

Servicing Mission 4 was most probably the last manned mission to Hubble, and with relatively minor exceptions, the goals of STS-125 were unchanged from the flight Sean O'Keefe cancelled in 2004. The seven crewmembers included: Scott D. Altman, commander; Gregory C. Johnson, pilot; Michael T. Good, MS1; K. Megan McArthur, MS2; John M. Grunsfeld,

MS3/payload commander; Michael J. Massimino, MS4; and Andrew J. Feustel, MS5. The destination was familiar territory to three of the crew: Altman and Massimino were making their second trip to Hubble and Grunsfeld his third. Two pair of astronauts – Grunsfeld/Feustel and Good/Massimino – made five spacewalks while McArthur operated the robotic arm.

STS-125 was the 126th flight of the Space Shuttle Program (STS-126 to the International Space Station had already flown, out of numerical sequence), marked the 30th flight of *Atlantis*, and was the last scheduled mission that did not directly support the ISS. It was also the first mission since the *Columbia* accident that did not fly to the ISS.

The payload bay of *Atlantis* contained several pallets that held equipment to service the telescope. The super lightweight interchangeable carrier (SLIC) housed the 980-pound (445-kilogram) Wide Field Camera 3 and two 475-pound (214-kilogram) battery modules. The SLIC weighed 1,750 pounds (empty) and was 15 feet (4.5 meters) wide and 8.6 feet (2.6 meters) long. The orbital replacement unit carrier (ORUC) contained the refurbished fine guidance sensor, rate sensor units (gyroscopes), and the Cosmic Origins Spectrograph. The flight support system (FSS) served as the berthing platform for Hubble and provided all electrical and mechanical interfaces between the Orbiter and the telescope while Hubble was docked. The FSS also contained

The STS-125 crew patch shows Hubble along with a representation of its many scientific discoveries. The overall structure and composition of the Universe is shown in blue and filled with planets, stars, and galaxies. The black background is indicative of the mysteries of dark-energy and dark-matter. The red border represents the red-shifted glow of the early Universe, and the limit of the Hubble's view into the cosmos. Soaring by the telescope is Space Shuttle, which initially deployed Hubble and has enabled astronauts to continually upgrade the telescope.

The crew immediately after the 2006 press briefing at the Johnson Space Center where the resurrected SM4 was announced. At right, is the official crew portrait. (All Courtesy of NASA)

Scott D. "Scooter" Altman

Gregory C. "Ray-J" Johnson

Michael T. "Bueno" Good

K. Megan McArthur

Trying on new shoes ...

John Mace Grunsfeld

Michael J. "Mass" Massimino

Andrew J. "Drew" Feustel

the soft capture mechanism that was attached to the aft end of the spacecraft to aid future visits to the telescope. The multi-use logistic equipment carrier (MULE) contained a number of components and spare parts for the telescope, including the relative navigation system, as well as extra cans of lithium hydroxide (LiOH) that could be used to scrub carbon dioxide from the Orbiter if the mission had to be extended for any length of time.

At the end of SM4, Hubble had a full set of scientific instruments in its light path for the first time since Thomas Akers and Kathy Thornton pulled out the High Speed Photometer in December 1993 to make room for the Corrective Optics Space Telescope Axial Replacement that fixed the spherical aberration in the primary mirror. At the time, one instrument location had to be sacrificed to make room for the corrective mirrors, but all of the new instruments have corrective optics built-in.

Replacing COSTAR, SM4 installed the Cosmic Origins Spectrograph (COS), an extremely sensitive ultraviolet spectrograph that will enable astronomers to begin mapping the structure and composition of galaxies and intergalactic gas shaped in part by dark matter. Wide Field Camera 3 was installed in place of the earlier WFPC2, and will allow the Hubble to probe deeper into the Universe than ever before in wavelengths ranging from ultraviolet through visible into the near-infrared. The old Wide Field/Planetary Camera 2 was removed and returned to Earth. The crew also replaced parts of the spacecraft to ensure it would function at least five more years, providing astronomers with the most capable telescope to ever orbit Earth.

In the high bay of the Payload Hazardous Servicing Facility (PHSF) at the Kennedy Space Center, a worker removes the protective wrapping from the orbital replacement unit carrier (ORUC). The refurbished fine guidance sensor is already stowed in the carrier. (NASA)

In April 2008, the United States Postal Service announced a First Class stamp honoring Edwin Hubble, and the STS-125 crew participated in the promotional campaign. (NASA)

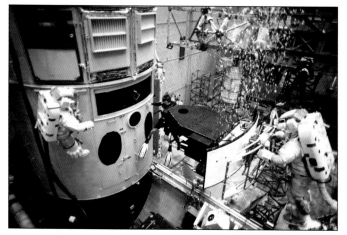

John Grunsfeld and Drew Feustel practice installing the Wide Field Camera 3 in the Neutral Buoyancy Laboratory at JSC. Note the safety divers in normal wetsuits and SCUBA gear. (NASA)

CREW TRAINING

NASA announced the crew for STS-125 in October 2006 and they began training soon afterward. As with previous Hubble servicing missions, the five planned extravehicular activities (EVA) dominated the mission-specific training. Most of the training was conducted at Johnson Space Center (JSC) with some training at other NASA centers, particularly Goddard, which is home to Hubble operations and engineering.

Some aspects of training for STS-125 were common in Space Shuttle missions in the pre-ISS era but had become unique since the *Columbia* accident. The low inclination and high altitude of Hubble's orbit resulted in different abort boundaries during ascent, and contingency abort landing options in the event of multiple engine failures were more limited. The track and capture of the free-flying Hubble by the remote manipulator system (RMS) on *Endeavour* was the first time this operation had been performed since the previous Hubble mission in 2002, and only the second time it had been performed for any payload since the capture of the SPARTAN spacecraft by STS-95 in 1998.

Training is conducted in a variety of facilities. Individual topics are usually introduced in classroom briefings. The single system trainers (SST) are simple flight-deck mockup simulators used extensively to train astronaut candidates in systems pro-

cedures and malfunction response, but in flight-specific training they are principally used for proficiency and qualification training. Once the crew has completed qualification in the necessary areas, they begin training as a crew in the shuttle mission simulator (SMS). The SMS is a complex facility capable of simulating all phases of a Space Shuttle flight, and consists of two fixed-base simulators and a motion-base simulator. The motion base simulator contains a mockup of the Orbiter forward flight deck mounted on a set of hydraulic actuators to simulate the dynamic motion of the Orbiter during launch, ascent, entry, and landing. The fixed base simulators contain mockups of the entire Orbiter flight deck and parts of the mid-deck, and are used to train the orbital phases of flight.

SMS training spans two distinct training flows. Prior to the delivery of a flight-specific simulator load, crews receive "flight-similar" training using a load from a previous flight. This presented a challenge for STS-125, since none of the recent flights (all to ISS) were similar, and the previous Hubble flight, STS-109, was so far in the past that intervening upgrades to the simulator and the Orbiter's flight software had rendered this load incompatible with the SMS. An interim load was improvised using updated Orbiter flight software along with selected simulator models from the STS-109 load, and the final STS-125 load was delivered on an accelerated schedule.

Other facilities are used to train specialized tasks required during the flight. The systems engineering simulator (SES) contains a mockup of the aft flight deck and is used to train rendezvous, proximity operations, and robotics. The dynamic skills trainer (DST) is a desktop simulator based on the same software as the SES and is used for robotics training and for proximity operations proficiency. The virtual reality (VR) laboratory uses head-mounted visual systems to train EVA techniques.

The most visually impressive training facilities are the Neutral Buoyancy Laboratory (NBL) and the Space Vehicle Mockup Facility (SVMF). The NBL is the world's largest swimming pool and is used for EVA training. The SVMF contains several orbiter mockups, including a full fuselage trainer and two crew compartment trainers. It is used for training in crew systems, photo/TV, emergency egress, and other areas.

Scott Altman (front seat) and Mike Massimino taxi out for a training flight in a Northrop T-38 Talon at Ellington Field near JSC. (NASA)

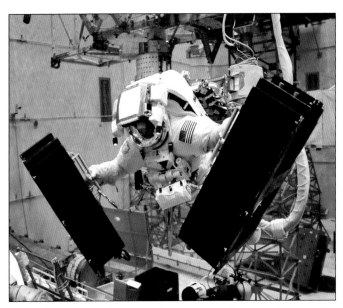

Megan McArthur sits at one of the consoles in the simulation control area of the Neutral Buoyancy Laboratory (NBL) at the Sonny Carter Training Facility (SCTF) near Johnson Space Center (JSC). Note the model of an Orbiter and Hubble. (NASA)

Drew Feustel practices installing new battery modules in the Neutral Buoyancy Laboratory. Feustel is handing an old battery module to John Grunsfeld while preparing to take a new battery module to the telescope. (NASA)

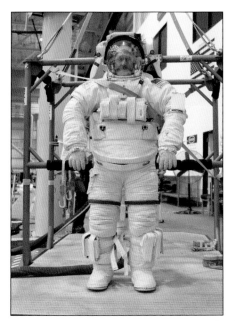

John Grunsfeld and Drew Feustel (partially obscured behind Grunsfeld) going for a swim in the NBL. Grunsfeld and Feustel are attired in training versions of the extravehicular mobility unit (EMU) spacesuit. The pair is lowered into the pool using a crane since the spacesuits are heavy and cumbersome. The astronauts trained extensively in the NBL to prepare them for their work on Hubble. (NASA)

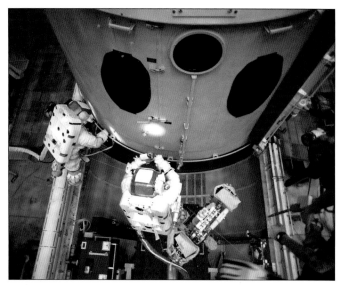

Two astronauts maneuver around the outside the Hubble mockup in the NBL. The two dark oval openings are for the star trackers that Hubble uses to aim itself. Visible at the bottom center is the mockup of the Orbiter payload bay, and at bottom right are two of the divers who conduct EVA training in the facility. (NASA)

Seated at the commander's station on the flight deck, Scott Altman participates in a post insertion/de-orbit training session in the crew compartment trainer (CCT-2) in the Space Vehicle Mockup Facility at JSC. Altman is wearing a training version of his advanced crew escape suit (ACES). (NASA)

STS-125 crewmembers practice Hubble retrieval in a mockup of the Orbiter aft flight deck in the systems engineering simulator (SES) in Building 16 at JSC. At bottom right is the display for the rendezvous and prox ops program (RPOP), a laptop computer program used during Hubble retrieval and ISS rendezvous and docking. (NASA)

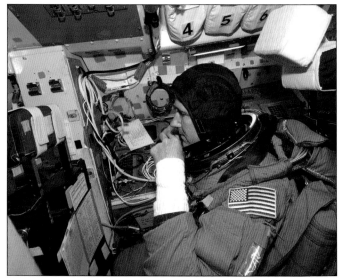

Michael Good adjusts his microphone during a training session in the crew compartment trainer (CCT-2). He is seated on the aft flight deck behind the pilot's seat. Once on-orbit, the mission specialist seats are folded and stowed to allow the aft flight deck to be used for rendezvous, RMS, and payload operations. (NASA)

John Grunsfeld practices crew bailout procedures in the crew compartment trainer in the Space Vehicle Mockup Facility at JSC. The bailout pole protrudes from the side hatch in the crew cabin and ensures the crew will clear the Orbiter's left wing as they bail out during an emergency. (NASA)

While seated on the middeck, Drew Feustel (foreground), Mike Massimino, and John Grunsfeld take a moment for a photo during a training session in one of the full-scale trainers in the Space Vehicle Mockup Facility at JSC. The crewmembers are attired in training versions of their advanced crew escape suits (ACES). (NASA)

The STS-125 flight control team participates in an integrated simulation in the White Flight Control Room in the Mission Control Center at JSC. The crew participates from the shuttle mission simulator (SMS). Note the simulated display of the remote manipulator arm on the projector screen at the front of the room. (NASA/James Blair)

On the Shuttle Landing Facility at the Kennedy Space Center, pilot Greg Johnson taxis one of the shuttle training aircraft (STA) toward the runway for practice landing the Orbiter. The STA is a Gulfstream II business jet that was modified to simulate the Orbiter's handling qualities during the atmospheric descent trajectory from approximately 35,000 feet altitude to landing on a runway. The left seat includes a full set of Orbiter displays and controls, while the right seat has the normal Gulfstream cockpit for the safety pilot. (NASA/Kim Shiflett)

The STS-125 crewmembers practice driving an M-113 armored personnel carrier as part of their emergency egress procedures during the terminal countdown demonstration test (TCDT). An M-113 will be available to transport the crew to safety in the event of a contingency on the pad before launch; it provides a relatively fire and blast-proof way to rapidly escape from the area if needed. At left the crew listens to an instructor, while the center photo shows Megan McArthur driving and the right shows John Grunsfeld at the controls. (NASA/Kim Shiflett)

During the TCDT, the craw also received instructions about using the slidewire baskets for emergency escape from the 195-foot level of the fixed service structure. In the left photo are Megan McArthur, Scott Altman, Greg Johnson, and Mike Massimino on the left side of the basket, and John Grunsfeld and Andrew Feustel on the right. In the center photo are Megan McArthur and Michael Good practicing climbing into the slidewire basket. The right photo shows where the baskets stop at the perimeter of the launch pad near a blast bunker. (NASA/Kim Shiflett)

The crew walks to the Astrovan after suiting up in the Operations and Checkout Building at KSC. The crew will participate in a simulated launch countdown as part of the TCDT. (NASA/Kim Shiflett)

From the front are Feustel, Massimino, and Grunsfeld on the middeck of Atlantis during the simulated launch countdown on 24 September 2008. The astronauts are wearing their ACES suits. (NASA/Kim Shiflett)

Scott Altman reads a checklist in the commander's seat of Atlantis. Despite being a large vehicle, the Orbiter crew compartment is cramped when the seats are erected and seven people are aboard. (NASA/Kim Shiflett)

After the simulated launch countdown, the crew practiced an escape by egressing toward the slidewire baskets. As is standard procedure, the crew climbed into the baskets, but did not ride the them down the slidewires to the bunkers. (NASA/Kim Shiflett)

KSC PROCESSING

The changes made to the External Tanks to minimize foam shedding during launch and ascent proved more difficult than expected to manufacture. Because the Michoud Assembly Facility (MAF) was unable to produce tanks as quickly as expected, it became necessary for NASA to further delay STS-125. Although the tank (ET-127) for the Hubble mission was ready in time for the September launch, the tank (ET-129) needed for the contingency launch-on-need (LON) flight of STS-400 was not, prompting a delay to 8 October. Ultimately, ET-127 for STS-125 arrived at KSC on 15 July, and ET-129 for STS-126/400 arrived on 11 August 2008. Like all External Tanks, both were transported on the *Pegasus* barge, towed by one of the SRB retrieval ships.

Further complicating preparations, Tropical Storm Fay pounded the central Florida coast for most of the week of 18 August, forcing NASA to close the Kennedy Space Center for several days and delaying the launch to 10 October.

The launch was again delayed after Hurricane Ike forced NASA to shut down the Johnson Space Center for over a week in mid-September. *Atlantis* had already been moved to LC-39A, and while JSC was closed *Endeavour* was moved to LC-39B and KSC processing proceeded normally. This was the first time since July 2001 that vehicles had been on both Space Shuttle launch pads at the same time, and, at the time, was expected to be the last ; as it turned out, there would be one more photo opportunity with Space Shuttles on both pads at the same time.

However, the closure of JSC adversely affected crew training for both STS-125 and STS-400, and also flight controller training for the two flights. In response, launch was delayed until 14 October. Another potential issue arose in mid-September when technicians discovered that insulation around the new batteries appeared to have come loose inside its packaging. Fortunately, fixing the problem did not further delay the launch.

One of the frustums is lowered onto a stand after being lifted from its transporter in the VAB transfer aisle. The frustrum forms the top of one of the Solid Rocket Boosters (SRB). (NASA/Jack Pfaller)

The two SRBs rise off the Mobile Launch Platform in the VAB. Later, the External Tank (ET) will be mounted between the SRBs and then Atlantis will be mated to the ET. (NASA/Jack Pfaller)

The External Tank (ET) for STS-125 arrived at the Kennedy Space Center on 15 July 2008. Like all tanks, ET-127 was towed from the Michoud Assembly Facility in Louisiana to Port Canaveral on the Pegasus barge by one of the two solid rocket booster retrieval ships (Freedom Star, in this case). Once at the port, a commercial tug boat takes over and tows the Pegasus to the turn basin near the Vehicle Assembly Building. (NASA/Jack Pfaller)

An unexpected concern for the Hubble mission came during the launch of STS-124 on 31 May 2008 when fire bricks separated from a 75- by 20-foot section of the east wall of the flame trench. These bricks had been installed when the pad was originally constructed during the 1960s to protect the concrete from the heat generated by a launch vehicle. Many of the bricks flew far enough that they landed in the estuary near the pad, and others impacted the perimeter fence around LC-39A, largely destroying it. Fortunately, none of the bricks hit Discovery during launch. There was also damage to the pad surface. Subsequent investigation determined that the age of the pad was largely to blame. (NASA/Kim Shiflett)

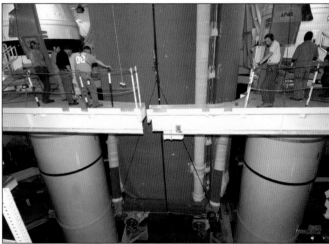

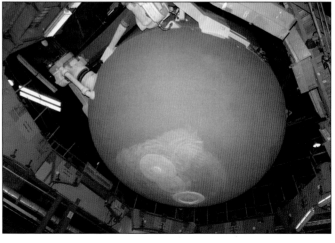

The first major step in preparing a Space Shuttle for launch is stacking the two Solid Rocket Boosters (SRB) on a mobile launch platform (MLP). For STS-125, this took place in high bay No. 3 of the VAB using MLP-1. After the white SRBs were stacked, the ET was slowly lowered between them on 3 August 2008. (NASA/Jack Pfaller)

The payload bay of Atlantis *before any payloads or carriers had been installed. The remote manipulator system arm is mounted on the port longeron (on the right of this photo), while the orbiter boom sensor system (OBSS) is on the starboard longeron.* (NASA/VITT via John Grunsfeld)

In the Orbiter Processing Facility (OPF) high-bay 1, technicians coordinate the movement of one of the three main engines being installed on Atlantis. Main engine No. 1 (the top-center of the three) has already been installed and the technicians are installing No. 2 in the right-hand position. The orbital maneuvering engines are behind the protective red covers, and the reaction control system thrusters are just outside the OMS engines. (NASA/Kim Shiflett)

Atlantis was rolled over from Orbiter Processing Facility 1 to the Vehicle Assembly Building on 22 August 2008. Carried by its wheeled transporter, the Orbiter arrived in the VAB at 11:05 pm. (NASA/Jack Pfaller)

In the Vehicle Assembly Building transfer aisle, workers detach Atlantis from its transporter. An overhead crane will lift Atlantis vertical and transfer the Orbiter to high bay 3 to be mated to its External Tank and Solid Rocket Boosters. The Orbiter is attached to the transporter in the same locations that mate with the External Tank. (NASA/Jack Pfaller)

Atlantis *hanging vertical in the transfer aisle. Note how close the vertical stabilizer is to the floor; the crane operators are very good. In the photo at right, the two umbilical doors can be seen surrounded by protective red pads.* (NASA/Dimitri Gerondidakis)

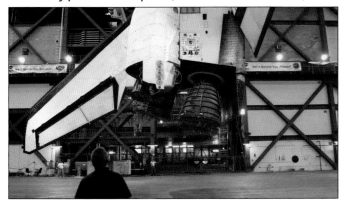

Atlantis is moved across the I-beam from the transfer aisle toward the waiting External Tank and Solid Rocket Boosters in high bay 3. Atlantis will be lowered and mated to the tank-booster stack on the mobile launcher platform. (NASA/Dimitri Gerondidakis)

A technician closely watches the progress of Atlantis *as it is lowered alongside the External Tank in high bay 3. The T-0 umbilical is immediately above the main engine; there is one on each side of the Orbiter. These umbilicals carry all electrical and data connections, and also the gaseous and cryogenic commodities needed to purge and load the External Tank.* (NASA/Dimitri Gerondidakis)

Atlantis on the crawlerway heading toward LC-39A. The stack is mounted to its mobile launch platform and is carried by the tracked crawler-transporter at a top speed of 0.9 mph. (NASA/Kim Shiflett)

Atlantis was rolled-out, on top of its mobile launch platform (MLP), from the VAB to LC-39A on 4 September 2008 beginning at 9:19 am. This is the view from the Launch Control Center (LCC) as the stack moved past at 0.9 mile per hour. An earlier 2 September roll-out date had been postponed due to Tropical Storm Hanna. (NASA/Kim Shiflett)

Above: *The tail service mast on the MLP contains the umbilicals that carry data, power, gasses, and propellants to the stack through the Orbiter aft fuselage. These connect to the umbilical plates shown on page 58.* (NASA/Kim Shiflett) Facing Page: *Highlighted by a rainbow after a afternoon Florida shower, Atlantis sits on LC-39A (foreground) and Endeavour on LC-39B.* (NASA/Troy Cryder)

The Wide Field Camera 3 (WFC3) shipping container is unloaded from a truck outside the Payload Hazardous Servicing Facility (PHSF) at KSC. In the PHSF, an overhead crane lifts the WFC3 and moves it to a work stand so that technicians can inspect the camera for shipping damage and prepare it for STS-125. (NASA/Jack Pfaller)

In the PHSF, the soft capture mechanism (SCM), part of the soft capture and rendezvous system (SCRS) is unpacked, inspected, and prepared for transfer to the multi-use lightweight equipment (MULE) carrier. The ring-like device attaches to Hubble's aft bulkhead and greatly enhances the current capture interfaces on Hubble, therefore significantly reducing the rendezvous and capture design complexities associated with the disposal mission. At right, technicians check one of the SCM sensors. (NASA/Troy Cryder)

Goddard engineers Richard Strafella (left) and Larry Dell hold a new outer blanket layer (NOBL) that are being used to replace external insulation blankets. Astronauts will install this particular NOBL on bay 8. At right is the logo used by the Goddard Space Flight Center for SM4. (NASA)

Michael Good holds a rate sensor unit during a crew equipment interface test (CEIT), which provides experience handling tools, equipment, and hardware. A CEIT is held at KSC prior to every mission. Also visible are Drew Feustel, Mike Massimino, and John Grunsfeld. (NASA/Kim Shiflett)

Above and Right: *The flight support system (FSS) carried the soft capture mechanism and was the device that held Hubble in the payload bay while astronauts worked on the telescope. The FSS has been used on every servicing mission, but this is the first time it held the SCM, which was left attached to Hubble to aid in the future docking of a yet-to-be-designed deorbit module.* (NASA/Troy Cryder)

Above: *The box containing the Cosmic Origins Spectrograph on the orbital replacement unit carrier.* Left: *The super lightweight interchangeable carrier (SLIC) that housed the 980-pound Wide Field Camera 3 and two 475-pound battery modules.* (NASA/Troy Cryder)

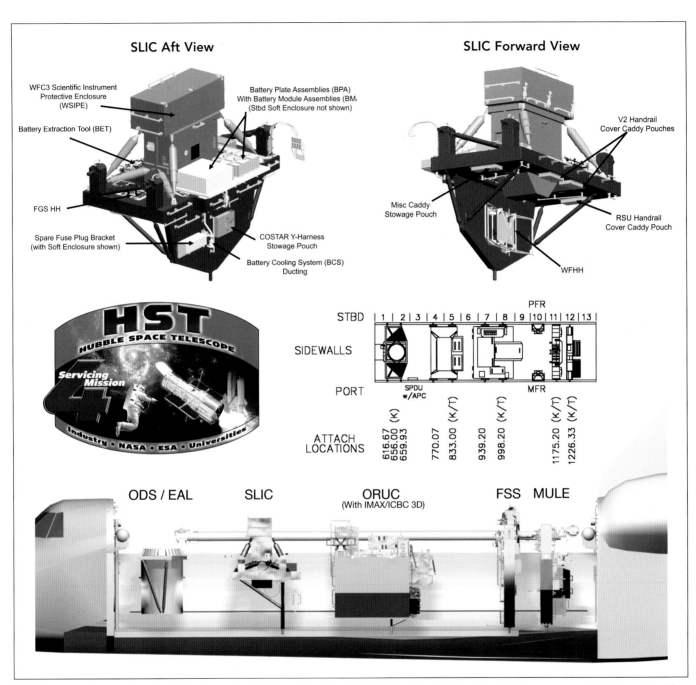

SLIC Aft View

WFC3 Scientific Instrument Protective Enclosure (WSIPE)

Battery Extraction Tool (BET)

Battery Plate Assemblies (BPA) With Battery Module Assemblies (BM) (Stbd Soft Enclosure not shown)

FGS HH

Spare Fuse Plug Bracket (with Soft Enclosure shown)

COSTAR Y-Harness Stowage Pouch

Battery Cooling System (BCS) Ducting

SLIC Forward View

V2 Handrail Cover Caddy Pouches

Misc Caddy Stowage Pouch

RSU Handrail Cover Caddy Pouch

WFHH

HST HUBBLE SPACE TELESCOPE
Servicing Mission 4
Industry · NASA · ESA · Universities

STBD	1	2	3	4	5	6	7	8	9	10	11	12	13

SIDEWALLS

PORT

SPDU w/APC

PFR

MFR

ATTACH LOCATIONS

616.67 (K)
656.00 (K)
659.93

770.07

833.00 (K/T)

939.20

998.20 (K/T)

1175.20 (K/T)

1226.33 (K/T)

ODS / EAL   SLIC   ORUC (With IMAX/ICBC 3D)   FSS   MULE

The diagrams above and on the facing page depict the payloads for Hubble Servicing Mission 4 and their arrangement in the payload bay of Atlantis. The orbiter docking system (ODS), with the docking mechanism removed, functions as the external airlock (EAL). Toward the rear of the payload bay, on the sidewalls just forward of the flight support system (FSS), are the portable foot restraint (PFR) and manipulator foot restraint (MFR) that are used to carry spacewalking astronauts on the end of the remote manipulator system (RMS). (NASA)

ORUC Aft View

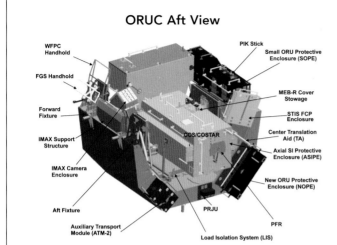

WFPC Handhold
FGS Handhold
Forward Fixture
IMAX Support Structure
IMAX Camera Enclosure
Aft Fixture
Auxiliary Transport Module (ATM-2)
PIK Stick
Small ORU Protective Enclosure (SOPE)
MEB-R Cover Stowage
STIS FCP Enclosure
COS/COSTAR
Center Translation Aid (TA)
Axial SI Protective Enclosure (ASIPE)
New ORU Protective Enclosure (NOPE)
PRJU
Load Isolation System (LIS)
PFR

ORUC Forward View

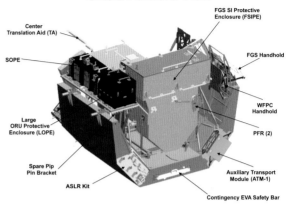

Center Translation Aid (TA)
SOPE
Large ORU Protective Enclosure (LOPE)
Spare Pip Pin Bracket
ASLR Kit
FGS SI Protective Enclosure (FSIPE)
FGS Handhold
WFPC Handhold
PFR (2)
Auxiliary Transport Module (ATM-1)
Contingency EVA Safety Bar

FSS Aft View

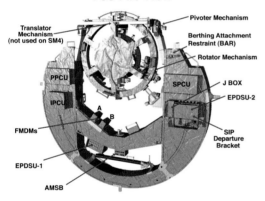

Translator Mechanism (not used on SM4)
PPCU
IPCU
FMDMs
EPDSU-1
AMSB
Pivoter Mechanism
Berthing Attachment Restraint (BAR)
Rotator Mechanism
SPCU
A
B
J BOX
EPDSU-2
SIP Departure Bracket

FSS Forward View

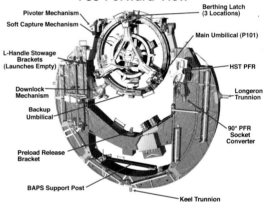

Pivoter Mechanism
Soft Capture Mechanism
L-Handle Stowage Brackets (Launches Empty)
Downlock Mechanism
Backup Umbilical
Preload Release Bracket
BAPS Support Post
Berthing Latch (3 Locations)
Main Umbilical (P101)
HST PFR
Longeron Trunnion
90° PFR Socket Converter
Keel Trunnion

MULE Aft View

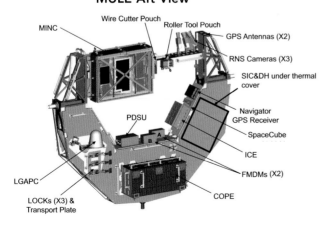

MINC
Wire Cutter Pouch
Roller Tool Pouch
GPS Antennas (X2)
RNS Cameras (X3)
SIC&DH under thermal cover
Navigator GPS Receiver
SpaceCube
ICE
FMDMs (X2)
COPE
PDSU
A
B
LGAPC
LOCKs (X3) & Transport Plate

MULE Forward View

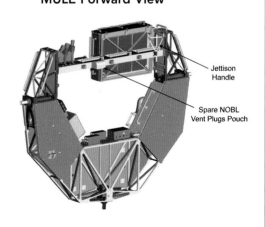

Jettison Handle
Spare NOBL Vent Plugs Pouch

There are two methods of installing payloads into an Orbiter. One is to install the payloads while the vehicle is horizontal in the Orbiter Processing Facility (OPF) before it is stacked. The other is to install the payload while the Orbiter is vertical, after it has been moved to the launch pad. SM4 used the latter. To accomplish this, the payloads are first installed in a special canister that is the exact size and shape of the payload bay. Here, the payload canister is moved from the Payload Hazardous Servicing Facility (PHSF) to the Canister Rotation Facility (CRF). The canister will be transferred to LC-39A and the payload will be loaded into Atlantis' payload bay. (NASA/Jack Pfaller)

Once the canister arrives at LC-39A, the rotating service structure (RSS, on the left side of the left photo) is rolled away from the Orbiter, the canister slides upward, and the payloads are transferred into the payload changeout room (PCR). The canister is then lowered and removed from the launch pad. The RSS rolls forward and the payloads are transferred are transferred into the Orbiter. Unfortunately, on the morning of 21 September 2008, when technicians went to raise the canister into the PCR, they found that teflon-covered shoes that help the canister move along guide rails did not fit properly. The shoes were removed (top right photo) and the teflon coating was shaved slightly. (NASA)

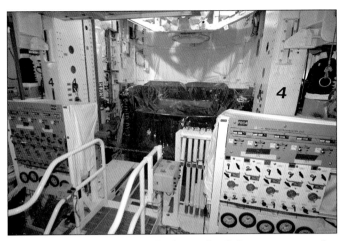

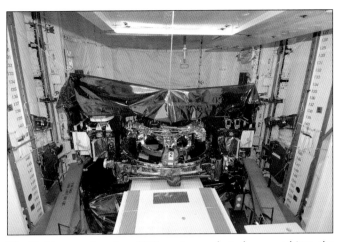

One of the control areas inside the payload changeout room that controls the payload ground-handling mechanism (PGHM). Since the launch complex was built in the early 1970s, most of the controls are manual with little automation. (NASA/Jack Pfaller)

The flight support system (FSS) carrier ready to be moved into the rear part of the payload bay. The protective wrapping stays on until everything is transfered into the Orbiter and the payload bay doors are ready to be closed. (NASA/Jack Pfaller)

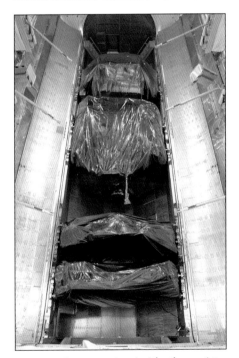

In protective wrapping inside the canister are (from top): the multi-use logistic equipment carrier, flight support system, orbital replacement unit carrier, and the super lightweight interchangeable carrier. (NASA)

The carriers will be transferred into the payload changeout room using the payload ground-handling mechanism, viewed here from the rear as it transfers the payload into the Orbiter. (NASA/Jack Pfaller)

The super lightweight interchangeable carrier (at bottom) is installed behind Atlantis' external airlock in the payload bay. The access door on the airlock is on the bottom in this photo. (NASA/Jack Pfaller)

HUBBLE IN TROUBLE STS-125 ON HOLD

Two weeks before the scheduled 14 October 2008 launch of STS-125, the Hubble Space Telescope suffered a major problem when the primary science computer shut itself down. The science instrument control & data handling (SIC&DH) unit transfers data from the science instruments to the ground and ground commands to the instruments. The SIC&DH consists of a NASA Standard Spacecraft Computer, two data formatter units, two central processing units, and other hardware. It is not related to the primary spacecraft control computers, and Hubble itself was never in any danger, but all science operations ceased.

Ironically, the failure may have happened at the most opportune time. Dr. Edward J. Weiler, NASA Associate Administrator for the Science Mission Directorate, commented, "Think about if

Atlantis at LC-39A on 15 October 2008 waiting for the winds to calm enough for the payload to be removed. The payload canister is in position on the rotating service structure at the payload changeout room to receive the Hubble hardware. (NASA/Tim Jacobs)

this failure had occurred two weeks after the servicing mission, we had put two brand new instruments in and thought we extended the lifetime for 5, 10 years, and this thing failed ..."

NASA immediately postponed STS-125 until engineers could ascertain the cause of the shutdown and develop a plan to correct it. Each month's delay in STS-125 reportedly cost $10 million. Initially, NASA thought it might be able to launch in January 2009, then the date was tentatively set for late February. It would soon become obvious that was optimistic.

To allow other Space Shuttle missions to proceed, the Hubble payload was removed from *Atlantis* on 15 October and the Orbiter was rolled-back from LC-39A to the VAB on 20 October. *Endeavour* was then moved from LC-39B where it had been waiting as the STS-400 rescue vehicle to LC-39A for launch as STS-126 to the International Space Station.

In the meantime, engineers at Goddard and the STScI began procedures to switch Hubble to a backup channel called the B-side. This backup channel had not been operated since the telescope had been launched, meaning it had sat idle for 18 years. As part of the switchover, the entire telescope was placed into safe mode, for only the sixth time since the observatory was launched in 1990.

The switchover was completed on 15 October. However, the effort was not completely successful since an unrelated power-supply problem prevented the Advanced Camera for Surveys from rebooting properly, and then the B-side computer failed. Engineers eventually successfully completed the switchover using elements of both the A-side and B-side, and the telescope began transmitting its typically dramatic photographs of the Universe, with the first science image, of an odd pair of galaxies called Arp 147, released on 30 October.

In the meantime, NASA was making plans to replace the failed SIC&DH unit on a rescheduled Servicing Mission 4. The unit was designed to be replaced on-orbit, and astronauts demonstrated they could replace it in about two hours, so no major impacts were expected to the servicing mission. A spare SIC&DH unit located at Goddard was brought out of storage and a long process begun to validate the unit and certify it for flight. Although the unit had been built as a flight spare, it had never gone through the entire qualification cycle, and over the years had been used in ground simulations.

Testing of the spare unit took until mid-December 2008, and electro-magnetic interference checks, vibration tests, and thermal vacuum chamber testing ran until early March 2009. After that, final acceptance testing took place at Goddard in late March, and the unit was delivered to KSC in early April.

Initially, mission managers had hoped to launch STS-125 in late February 2009, but by the end of October 2008 it became obvious that the testing and certification of the replacement hardware would push the launch until the end of April at the earliest. Problems with an obsolete component during early testing also delayed the effort.

The majority of the flight hardware, tools, and support equipment were stored at Kennedy. A small amount of new

work such as re-lubricating the latches on the soft capture Mechanism and testing the motors on the flight support system would be conducted at KSC. The new batteries to be installed during the mission were placed in cold storage at Goddard and were returned to KSC just prior to launch.

Once the decision had been made to postpone STS-125 even further, *Atlantis* was demated from the STS-125 External Tank and Solid Rocket Boosters, and that stack was used to support STS-119/15A with OV-103/*Discovery*. *Atlantis* was demated on 8 November and stored in OPF-3. During the downtime, several of the leading edge RCC panels on *Atlantis* were replaced as part of normal maintenance.

On 4 December, the launch of STS-125 was finally rescheduled for 12 May 2009, although that date was subsequently advanced to 11 May. During the first five months of 2009, the crew continued training, scientists and technicians readied the new science instruments and the repaired SIC&DH unit, and the ground crew processed *Atlantis* and *Endeavour* in preparation for STS-125 and its potential STS-400 rescue flight.

The payload canister being lowered onto its transporter after the Hubble hardware was transferred into it. The Orbiter can now be moved back to the VAB. (NASA/Kim Shiflett)

The payload canister was transferred back to the Payload Hazardous Servicing Facility for storage. The red umbilical lines connect the canister to air conditioning equipment on the transporter to keep the payload in a controlled environment. The multi-wheel transporter was built by KAMAG Transporttechnik GmbH in Germany. (NASA/Tim Jacobs)

Four of these giant mounts are what support the mobile launcher platform and Space Shuttle stack while it sits on the launch pad. The crawler-transporter is capable of lifting the MLP off these mounts when it carries it back and forth to the VAB. (NASA/Kim Shiflett)

Above, Below, and Right: **Atlantis** *began its trip back to the VAB at 06:48 am on 20 October 2008. On its way back to the VAB, the stack passed the Pegasus barge delivering the ET for STS-119 (below). The photo at upper right shows* Atlantis *arriving in high bay 3 of the Vehicle Assembly Building, where it was stored awaiting launch.* (NASA/Kim Shiflett)

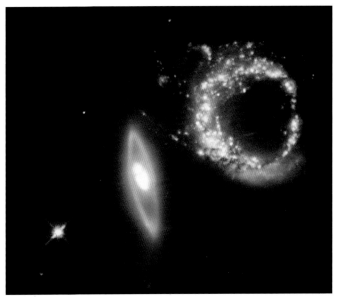

Above: *The first science image from Hubble after the trouble with the SIC&DH unit was this WFPC2 image of a pair of gravitationally interacting galaxies called Arp 147. The two galaxies happen to be oriented so that they appear to mark the number 10. The left-most galaxy, or the "one" in this image, is relatively undisturbed apart from a smooth ring of starlight. It appears nearly on edge to our line of sight. The right-most galaxy, resembling a "zero," exhibits a clumpy, blue ring of intense star formation. The blue ring was most probably formed after the galaxy on the left passed through the galaxy on the right. The dusty reddish knot at the lower left of the blue ring probably marks the location of the original nucleus of the galaxy that was hit. Arp 147 lies in the constellation Cetus, and it is more than 400 million light-years away from Earth. This composite was assembled from WFPC2 images taken with blue, visible-light, and infrared filters.* (STScI)

Left: Atlantis *in high bay 3 of the VAB. Note the myriad of work platforms that surround the Space Shuttle stack when it is in the VAB, allowing technicians access to all critical areas of the vehicle. The forward reaction control system (FRCS) module in the nose of the Orbiter contains thrusters that point in five directions (up, down, left, right, and forward) and, along with an RCS module in each aft-mounted orbital maneuvering system (OMS) pod, provides attitude control for the vehicle while it is on-orbit.* (NASA/Kim Shiflett)

The second stacking for STS-125 took a giant step forward on 15 January 2009 when ET-130 was lowered into place between the solid rocket boosters in VAB high bay 3. This stack used a different ET and SRBs than the original stack. The components from the original stack were used for STS-119. (NASA/Jack Pfaller)

Atlantis is suspended by a crane while its transporter is moved away in the transfer aisle of the VAB before it was lifted into high bay 3. Hardpoints built-into the major structural bulkheads at the front and back of the payload bay allow the Orbiter to be lifted. (NASA/Cory Huston)

A view seldom seen. The second STS-125 stack was moved from VAB high bay 1 to high bay 3 on 10 February 2009 to allow the STS-127 stack to use high bay 1. It is unusual to see this side of the External Tank without an Orbiter attached. (NASA/Tim Jacobs)

On 24 March, Atlantis was moved from the Orbiter Processing Facility to the Vehicle Assembly Building and mated with the already-mated External Tank and Solid Rocket Boosters. The top of the ET can be seen at the bottom of the photo. (NASA/Kim Shiflett)

Servicing the Hubble Space Telescope

Scott Altman flies a Gulfstream II shuttle training aircraft (STA) over White Sands Test Facility in New Mexico. The left seat of the STA is configured with a set of Orbiter displays and controls, including the heads-up display. (NASA/Richard N. Clark, AOD division chief)

The mission had been waiting for the replacement science instrument command and data handling (SIC&DH) unit to be completed by the Goddard Space Flight Center, and it finally arrived at KSC on 2 April. (NASA/Dimitri Gerondidakis)

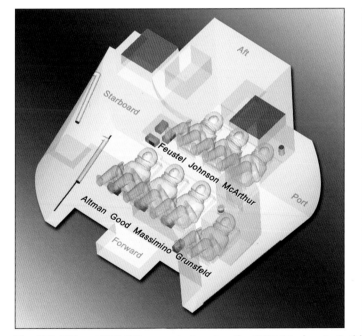

The STS-125 and STS-400 crews train using the CCT-I Mockup in Building 9NW at JSC. If the STS-125 crew would have been rescued on-orbit, all seven crewmembers would have returned to Earth on the middeck of Endeavour, as shown here. (NASA/Devin Boldt)

LAUNCH-ON-NEED (LON) – STS-400

STS-400 was the designation given to the Launch-on-Need (LON) mission that would have been launched to rescue the crew of *Atlantis* if the Orbiter had become disabled during STS-125. Due to the orbital inclination of Hubble, *Atlantis* was unable to use the International Space Station (ISS) as a "safe haven" in the event of catastrophic structural or mechanical damage. Unlike missions to the International Space Station, which can sustain a stranded Space Shuttle crew for up to 90 days, the STS-125 mission to Hubble offered no additional supplies other than those carried aboard the Orbiter (which carried 25 days worth of food and water). In the event of catastrophic damage to *Atlantis*, the crew would have conserved power and oxygen to extend life support for as much as 25 days while *Endeavour* was readied for launch. The two STS-400 missions were the first time since

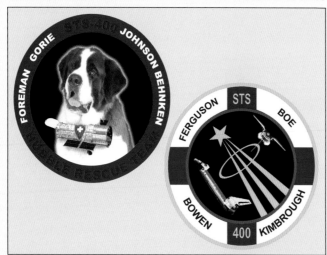

Unofficial patches for the first (top left) and second (lower right) STS-400 crews. (Upper: Michael Foreman; Lower: Tim Gagnon and Jorge Cartes / collectSPACE.com).

July 2001 – and probably the last time – that Space Shuttles simultaneously occupied both launch pads.

"The astronaut office was not unified as to the opinion of whether we need a launch-on-need shuttle," NASA Administrator Mike Griffin said when he approved the Hubble servicing mission. "Statistically, it would be silly to say anything other than we don't need a rescue mission because we have a much higher chance of losing a shuttle from some other cause for which a rescue mission can't help you." Griffin continued, "But there is a small chance that we could lose a shuttle, an event which could be remedied by having a launch-on-need capability. To protect against that small chance, we have decided that we will maintain a rescue capability for this mission." The capability to rescue a space shuttle crew stranded in space was not an explicit recommendation of the Columbia Accident Investigation Board (CAIB). However, as part of a "raising the bar" initiative, NASA decided to implement the ability to launch a rescue flight.

STS-125 presented unique challenges to providing a rescue capability, differing significantly from the standard ISS rescue scenario. The difference in orbital inclination between Hubble and ISS made it impossible for *Atlantis* to reach ISS in the event of an emergency, meaning the crew would have to survive on Orbiter consumables alone. This requires the LON mission to be ready to launch much more quickly than for an ISS mission. The lack of a docking mechanism on STS-125 (and inadequate clearances for safe docking even if one were present) required that *Endeavour* grapple *Atlantis* using the remote manipulator system and transfer the stranded crew via extravehicular activity (EVA). This was not an activity performed on ISS missions and therefore required unique training. This ruled out the use of the next ISS mission (STS-126) crew as the Hubble LON crew because their schedule would not allow training for both missions at once.

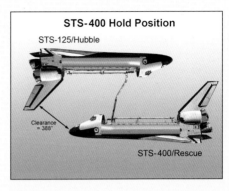

STS-400 Hold Position

STS-125/Hubble

Clearance = 388"

STS-400/Rescue

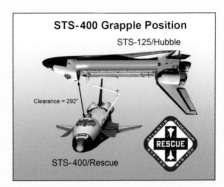

STS-400 Grapple Position

STS-125/Hubble

Clearance = 292"

STS-400/Rescue

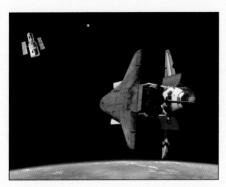

To minimize training requirements, NASA decided to assign the nucleus of a recently flown crew to the Hubble LON mission. In June 2008, NASA assigned commander Dominic L. Pudwill Gorie, pilot Gregory H. Johnson, and mission specialists Robert L. Behnken and Michael J. Foreman as the STS-400 crew. They had previously flown on STS-123 in March 2008, which delivered the first component of the Kibo Japanese Experiment Module (JEM), to ISS along with the Canadian Special Purpose Dexterous Manipulator (SPDM) Dextre.

When STS-125 was delayed for several months due to the SIC&DH failure on Hubble, a new STS-400 crew was assigned from the nucleus of the recently flown STS-126 crew: commander Christopher J. Ferguson, pilot Eric A. Boe, and mission specialists Robert S. "Shane" Kimbrough and Stephen G. Bowen.

Following launch, the *Endeavour* crew would have spent the rest of the day checking out rendezvous equipment and the remote manipulator arm. *Endeavour* would have rendezvoused from beneath *Atlantis* approximately 23 hours after launch with Shane Kimbrough using the robot arm to grapple a fixture on the forward end of the orbiter boom sensor system mounted along the right side of *Atlantis'* payload bay. The grapple would have occurred with the Orbiters perpendicular to one another for clearance; the distance between the two at grapple would be 35 feet. After capture, a 90-degree yaw of the arm would have put the Orbiters payload-bay-to-payload-bay. This would have provided the most stable attitude and position ahead of the three spacewalks needed to relocate *Atlantis'* crew to *Endeavour*.

The third day of the mission would have been dedicated to setting up the translation path from *Atlantis* to *Endeavour* and the start of crew transfer. John Grunsfeld and Andrew Feustel would string a cable along the length of the remote manipulator arm to serve as the translation path and Megan McArthur would move to *Endeavour*, followed by Feustel and Grunsfeld who would have spent the night on *Endeavour*. The first spacewalk was scheduled to last 4 hours, 50 minutes.

On the fourth day of the flight, Grunsfeld would have suited up and returned to *Atlantis* and assist Mike Massimino and Gregory Johnson. The second spacewalk was planned to last just under 2 hours, leaving Grunsfeld and Johnson on *Endeavour*. The third spacewalk would have taken place the same day with Massimino moving back to *Atlantis* to assist Michael Good and Scott Altman. This spacewalk was budgeted for about 2 hours, 30 minutes. Prior to leaving *Atlantis*, Altman would reconfigure the vehicle so that flight controllers could command a destructive entry, with the Orbiter burning up over the Pacific Ocean. With all seven STS-125 crewmembers aboard, Shane Kimbrough would have released the grapple fixture on *Atlantis'* boom ending the rescue operation.

The following day would focus on conducting a survey to ensure OV-105 was not damaged during ascent or the on-orbit operations using the remote manipulator arm and the orbiter boom sensor system. The "new" crewmembers from *Atlantis* would have performed the checkout. On the sixth day of the mission, the crew would have focused on cabin stowage in preparation for return home. Routine day-before-landing activities and systems checkouts would have taken place on the seventh day, with deorbit and landing on the eighth day.

The day after STS-125 was launched, KSC personnel began preparing Endeavour for launch should the on-orbit inspection show Atlantis was damaged. After a thorough assessment of the data from the late inspection, NASA cleared *Atlantis* for entry. Once *Endeavour* was released from its STS-400 rescue obligation, it was relocated from LC-39B to LC-39A for launch on the STS-127 mission to resupply the International Space Station on 13 June 2009.

FINALLY, LAUNCH

On 24 April 2009, NASA decided to move the launch of STS-125 forward one day to 11 May to allow the launch window to be expanded from 2 days to 3 days. The change was made official at the flight readiness review on 30 April. Everything was looking good as 11 May approached. *Atlantis* and *Endeavour* were both ready, and the crew was excited to finally undertake the mission they had been training for.

Dawn broke on a clear spring day in Central Florida, but as the appointed time approached, a single dark cloud seemed as if it would drift over the launch site. Fortunately, the cloud stalled over nearby Titusville, and *Atlantis* lifted-off on time, at 2:01 pm. Almost immediately after launch and during the ascent, flight systems reported problems with a transducer in one of the main engines and a circuit breaker; the crew was advised to disregard the resultant alarms and continue to orbit.

The post-launch inspection of the pad showed a 25-foot area on the north side of the flame deflector where some heat-resistant coating had come off. Fortunately, photo-analysis showed that none of the debris impacted the Orbiter during launch. An on-orbit inspection showed a 21-inch debris strike on the lower portion of the starboard leading edge chine, but the damage was not deep and after a careful analysis, mission managers decided the damage would not affect entry at the end of the mission. Both Solid Rocket Boosters were recovered without incident. The images of External Tank separation taken from cameras in the umbilical wells of the Orbiter could not be downlinked due to technical issues, but photos taken by the crew using handheld cameras showed an External Tank that was in excellent shape with very few areas where foam separated during ascent.

The crew arrived, again, at KSC in their T-38 trainers on 8 May. The non-pilot rated crewmembers fly in the back seat of the airplanes. From left are Megan McArthur, Michael Good, Greg Johnson, Scott Altman, John Grunsfeld, Mike Massimino, and Drew Feustel. (NASA/Jim Grossman)

An X-Band radar was installed on the Army LCU-2005, Brandy Station. The radar worked with a smaller X-band radar placed on the SRB retrieval ship Liberty Star to provide extremely high-resolution images of any debris that might be created during Atlantis' launch. (NASA/Kim Shiflett)

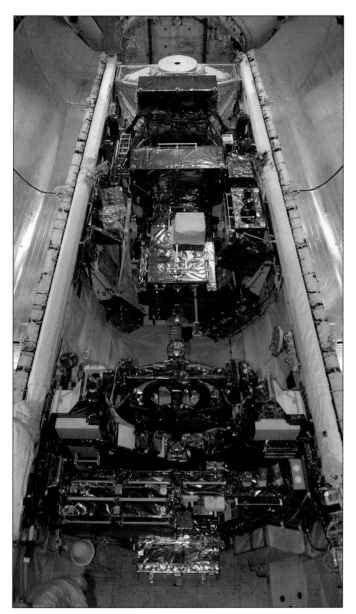

Workers close the right-hand payload bay door on Atlantis the weekend before launch. Partially concealed from view at the bottom is the flight support system (see photo at left). Note the radiator on the inside of the payload bay door behind the technicians at the right of the photo below. (NASA/Kim Shiflett)

The payload bay doors of Atlantis were closed on 8 May. At the bottom are the flight support system (FSS) with the soft capture Mechanism and the multi-use lightweight equipment carrier (MULE) with the science instrument command and data handling (SIC&DH) unit. At center is the orbital replacement unit carrier (ORUC) with the Cosmic Origins Spectrograph (COS), and an IMAX 3D camera. At top is the super lightweight interchangeable carrier (SLIC) with the Wide Field Camera 3 (WFC3). (NASA/Kim Shiflett)

Atlantis *as STS-125 on LC-39A (foreground) and* Endeavour *as STS-400 on LC-39B in late April 2009. Some significant changes had been made to Pad-B for the scheduled mid-summer Ares I-X demonstration flight for the Constellation Program (compare to photo on page 61). The most obvious is the three new lightning towers surrounding the pad. The Space Shuttle Program used a single lightning tower on top of the fixed service structure (FSS).* (NASA/Dimitri Gerondidakis)

Endeavour as STS-400 on LC-39B in the early morning of 17 April 2009. The pads are well lit, and can easily be seen from the nearby cities and by passengers on airliners transiting through Orlando. (NASA/Dimitri Gerondidakis)

Another try. Atlantis waiting on LC-39A on 17 April for its trip to Hubble. The 290-foot-high tower at right holds 300,000 gallons of water that is released just prior to the start of the main engines. The water flows through 7-foot-diameter pipes to a series of large "rainbirds" to control acoustic damage during launch. (NASA/Kim Shiflett)

Servicing the Hubble Space Telescope

Atlantis on LC-39A with the Vehicle Assembly Building (VAB) and Launch Control Center (LCC) in the background left. The rotating service structure is rolled-back from the Orbiter awaiting the payload canister that would install the Hubble repair parts. (NASA/Dimitri Gerondidakis)

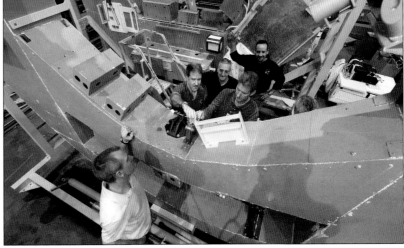

On 9 April 2009, Drew Feustel (bottom), John Grunsfeld (center left), Mike Massimino, and Michael Good (center, right) working with a Hubble Space Telescope mockup during a 1-g training session at the Johnson Space Center. (NASA/James Blair)

The night before launch, and all is well. The rotating service structure has been retracted, and cryogenic propellant loading commenced in the early hours of 11 May. The crew was asleep several miles south in the Operations and Checkout (O&C) building in the KSC industrial area. (NASA/Dimitri Gerondidakis)

The STS-125 crew head for the Astrovan outside the Operations and Checkout Building. In front of the Astrovan, from left, are Mike Massimino, Michael Good, Drew Feustel, John Grunsfeld, Megan McArthur, Gregory Johnson, and Scott Altman. (NASA/Kim Shiflett)

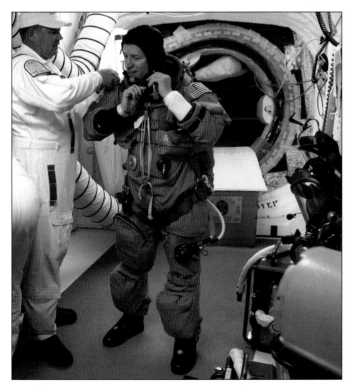

In the white room at the end of the orbiter access arm, Pilot Gregory Johnson is helped by the closeout crew putting on his harness, which includes a parachute pack, before crawling through the open hatch into Atlantis. (NASA/Sandra Joseph-Kevin O'Connell)

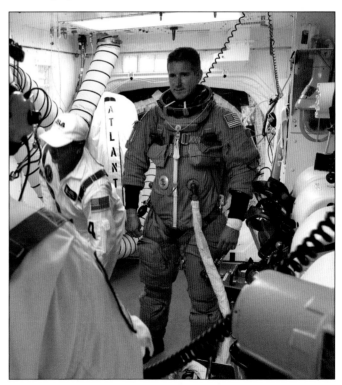

Michael Good prepares to enter Atlantis wearing his David Clark Company S1035 Advanced Crew Escape Suit (ACES). This full-pressure suit protects the astronauts against a loss of cabin pressure or in case they need to bail out. (NASA/Sandra Joseph-Kevin O'Connell)

Against a backdrop of clouds and framed by banks of trees, Atlantis roars off LC-39A on time at 2:01 pm EDT. Several hundred-thousand gallons of water are dumped into the flame trench to suppress the acoustic signature of the solid rocket booster and main engines, resulting in steam clouds as the vehicle lifts off. (This Page: NASA/Sandra Joseph-Kevin O'Connell; Facing Page: NASA/Tony Gray-Tom Farrar)

Servicing the Hubble Space Telescope

Atlantis *trails blue Mach diamonds in the exhaust of the three Space Shuttle Main Engines. The Mach diamonds are shock waves that form as the exhaust plume expands at high velocity.* (NASA/Michael Gayle-Rusty Backer)

Steam bellows out of the flame trench (above) and a fish-eye view (below) shows Atlantis as it lifts off the pad. Note the retracted white room in the lower left corner of the photo below. (Below: NASA/Tony Gray-Tom Farrar; all others on page: courtesy of Scott Andrews, Canon)

Atlantis is framed by the new lightning towers surrounding Endeavour on LC-39B. The STS-400 rescue mission would have been launched if Atlantis would have been irreparably damaged during its mission. (Photo courtesy of Scott Andrews, Canon)

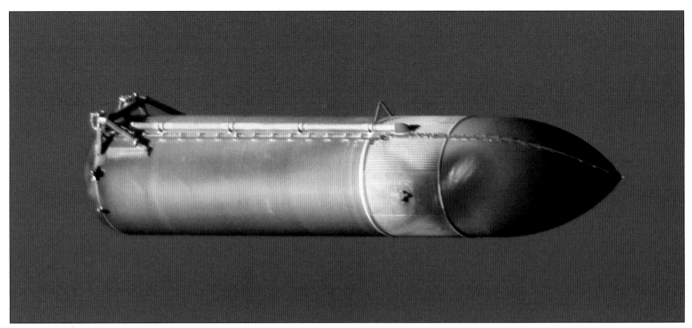

ET-130 after separation from Atlantis, as viewed by the crew. The scorching near the nose is from the SRB separation motors that push the Solid Rocket Boosters away from the stack when they separated several minutes earlier. Each ET is extensively photographed after separation to verify no insulating foam fell off as happened during the Columbia accident. Overall, this tank was in excellent condition, although several small pieces of foam were noted missing. Shortly after this photo was taken, the ET, as planned, began to tumble and ultimately broke up over the ocean. (NASA)

Unlike the External Tank, the Solid Rocket Boosters are recovered and reused after each mission. Each SRB is lowered by three large parachutes into the Atlantic Ocean about 141 miles east of Florida where two ships operated for NASA retrieve them. (NASA)

Each SRB is towed back to Hangar AF on the Cape Canaveral Air Force Station (CCAFS) adjacent to KSC where it is floated into a slip. The forward frustum and parachutes are also retrieved, returned to CCAFS, and refurbished for another mission. (NASA)

  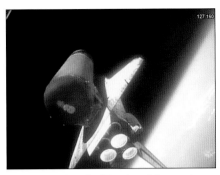

Each of the Solid Rocket Boosters is equipped with three relatively low-resolution video cameras – one in the aft skirt aimed forward, one in the frustrum aimed rearward, and one aimed at the External Tank attach point. Here are a few video captures that show scenes from the STS-125 launch. The photo at upper left is at main engine ignition (note the flame), then two during SRB separation, two showing the other SRB tumbling toward Earth, and one of the parachutes settling into the Atlantic Ocean after the booster splashed down. (NASA)

  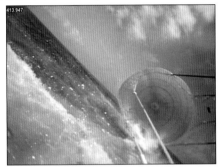

The left-hand SRB on the dock. Divers inserted a plug in the exhaust nozzle while the booster was bobbing in the Atlantic so that they could pump air into the booster to make it easier to tow back to base. Note the CEV capsule boilerplate in the background. (NASA)

The three large parachutes that lower the SRBs, plus the smaller parachutes used for the frustum, are all recovered and reused. The parachutes are wound onto large reels on the SRB retrieval ships, then transported to the Parachute Refurbishment Facility at KSC. (NASA)

REPAIRING HUBBLE

During the first day on-orbit, called Flight Day 2 (FD2, 12 May), the crew prepared the flight support system (FSS) for berthing, surveyed the wing leading edge reinforced carbon-carbon (RCC) for damage, and checked-out the extravehicular mobility units (EMU) that would be used during the spacewalks.

Atlantis grappled Hubble on FD3, but this did not go entirely as planned. NASA had difficulty establishing communications with Hubble through the Orbiter's payload interrogator (PI). The STOCC and crew could command Hubble via the PI, but only "in the blind." This caused delays in commanding and required Scott Altman to stationkeep around 150 feet from the telescope. Hubble's final roll maneuver to grapple attitude was also deleted, requiring Altman to yaw the Orbiter to allow RMS operator McArthur to reach the grapple fixture with the RMS. The stationkeeping and unexpected maneuvers consumed about 200 pounds of additional propellant. Post-grapple, it was determined that the command problems resulted from Hubble still being in science telemetry format rather than transitioning to the servicing format required by the Orbiter. The format was corrected and normal operations resumed.

The next five days were an intense flurry of extravehicular activity (EVA), with five spacewalks performed by alternating pairs of astronauts:

Flight Day 4, EVA1 (Grunsfeld/Feustel): Replaced the Wide Field Planetary Camera 2 (WFPC2) with Wide Field Camera 3 (WFC3); replaced the failed Science Instrument Control and Data Handling (SIC&DH) unit; installed the Soft Capture Mechanism (SCM) on the aft-end of the spacecraft.

Flight Day 5, EVA2 (Massimino/Good): Replaced the Rate Sensing Units (RSU) and bay 2 batteries.

Flight Day 6, EVA3 (Grunsfeld/Feustel): Replaced the Corrective Optics Space Telescope Axial Replacement (COSTAR) with the Cosmic Origins Spectrograph (COS); repaired the Advanced Camera for Surveys (ACS).

Flight Day 7, EVA4 (Massimino/Good): Repaired the Space Telescope Imaging Spectrograph (STIS); installed New Outer Blanket Layer (NOBL) on bay 8.

Flight Day 8, EVA5 (Grunsfeld/Feustel): Replaced bay 3 batteries and Fine Guidance Sensor 2 (FGS2), installed NOBL on bay 5.

John Grunsfeld uses a long lens to photograph Hubble through one of the upper windows on the aft flight deck while Atlantis was about 2,500 feet from the telescope. Shortly afterward, Scott Altman took manual control for the final approach. The window on the bulkhead behind Grunsfeld looks into the payload bay. (NASA)

Michael Good operated a handheld LIDAR (HHL) to determine the range and closing rate to Hubble as Altman performed the manual approach. Unlike the International Space Station (ISS), Hubble does not have laser retroreflectors on its surface, so the Trajectory Control Sensor (TCS) used for ISS approaches cannot be used. (NASA)

The joys of microgravity. Megan McArthur, Mike Massimino (center), and Drew Feustel, prepare to eat a meal on the middeck. The astronauts are allowed to select the meals from a fairly wide selection that are packed aboard the Orbiter. There is no "up" other than the general orientation of the items inside the crew cabin. (NASA)

Hubble secured to the flight support system (FSS) in the payload bay. The high-gain antennas (the black dish near the top) have been retracted flush against the sides of the telescope. (NASA)

An EVA checklist and cue cards arranged on the aft flight deck for EVA1. Spacewalks are choreographed and rehearsed down to the smallest detail, and the intravehicular activity (IVA) crewmember plays a key role in organizing and conducting the EVA. Mike Massimino and Michael Good were the IVA crewmembers for EVA1. (NASA)

John Grunsfeld (left) and Drew Feustel prepare for the mission's first spacewalk on Flight Day 4 (FD4), 14 May. This EVA replaced the Wide Field/Planetary Camera 2 (WFPC2) with the new Wide Field Camera 3 (WFC3) that will operate over a wider range of wavelengths with a larger field of view. (NASA)

John Grunsfeld wore a spacesuit with red stripes on the legs and corners of the backpack, while Drew Feustel used an all-white suit. This allowed the other crewmembers and flight controllers to tell the astronauts apart during their spacewalks. (NASA)

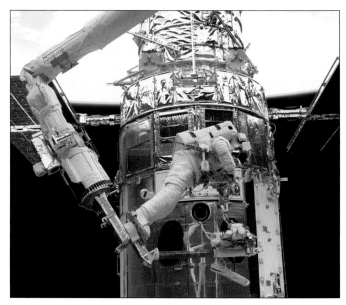

Drew Feustel is perched on the end of the Canadian-built remote manipulator system (RMS) arm, preparing to remove the WFPC2 during the first of five spacewalks. John Grunsfeld was working in the payload bay out of the camera's view. (NASA)

The WFPC2, which was located in the only radial bay that contains a science instrument, is the white horizontal rectangle in the middle of the telescope. The three black holes below WFPC2 are the star trackers, which are used to help determine the orientation of the telescope in space. (NASA)

Grunsfeld holds onto a handrail during the Flight Day 4 spacewalk. Hubble was designed to be maintained on-orbit and Lockheed incorporated 225 linear feet of handrails to assist the astronauts during servicing missions. Note the empty hole exposing the radial bay that will soon be filled by the WFC3. (NASA)

Grunsfeld removing tools from the tool stowage assembly (TSA) on the port side of the forward end of the payload bay during EVA1. The truss securing the external airlock in the payload bay contains a TSA on each side of the airlock. (NASA)

Feustel stands on the manipulator foot restraint (MFR) at the end of the remote manipulator arm. The WFC3 has been installed. The external differences between the old and new cameras are readily apparent when this photo is compared with the one at top left. (NASA)

Feustel holding the WFPC2, which was returned to Earth using the same location in the super lightweight interchangeable carrier (SLIC) that had contained WFC3. It must have been distracting to be working with such a gorgeous view of Earth below. (NASA)

Grunsfeld and Feustel secure WFPC2 to the SLIC in the payload bay for its return trip to Earth. Looking at the telescope, it is apparent that the WFC3 was fully installed before the astronauts secured the older WFPC2 in its carrier. (NASA)

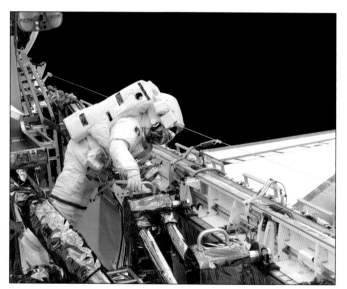

Grunsfeld at work in the payload bay during EVA1, photographed from the flight deck window (above) and by Drew Feustel (above right) while in a foot restraint at the end of the RMS, hence the overhead perspective. The line running above the sill is a EVA slidewire that provides a safety attachment point for the astronauts. (NASA)

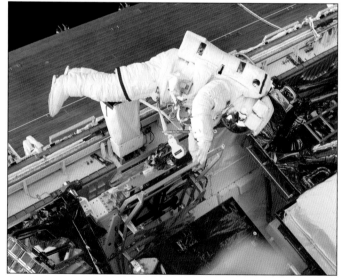

Facing Page: *This close-up of Grunsfeld was taken by Feustel, whose reflection is visible in the visor. The helmet is generally similar to that worn by Apollo astronauts and is made by Air-Lock, Inc., a subsidiary of the David Clark Company, a former underwear manufacturer most famous for modern aviation pressure suits.* (NASA)

The second spacewalk of the mission, on 15 May, was performed by Michael Good and Mike Massimino. The pair replaced the rate sensing units and the bay 2 batteries. Good wore red-and-white stripes on his spacesuit while Massimino used solid red stripes, like Grunsfeld (but the two were never out together). (NASA)

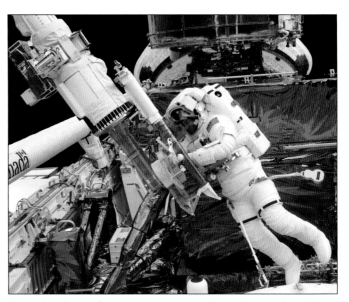

Note the dual safety tethers securing Massimino to the Orbiter. Unlike ISS EVAs, the Hubble spacewalkers did not have the Simplified Aid for EVA Rescue (SAFER) jetpacks on their suits; in case they became untethered, Altman would have rescued them using the ignominious-sounding "tool-chasing" procedure. (NASA)

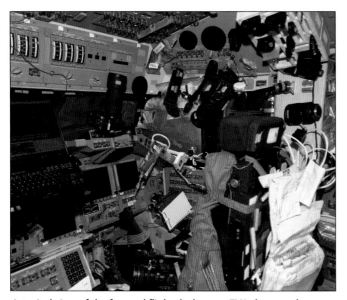

A typical view of the forward flight deck on an EVA day reveals a mess of cameras and lenses. Unlike wet film of yore, digital images may be downlinked to the ground almost immediately. The laptop computer at left displays the current position of Atlantis in its orbit. (NASA)

The solar panels on Hubble make unique window shades in this scene photographed from the aft flight deck. The larger "joy stick" controls the rotation of the Orbiter, while the smaller square controller is for the remote manipulator arm. (NASA)

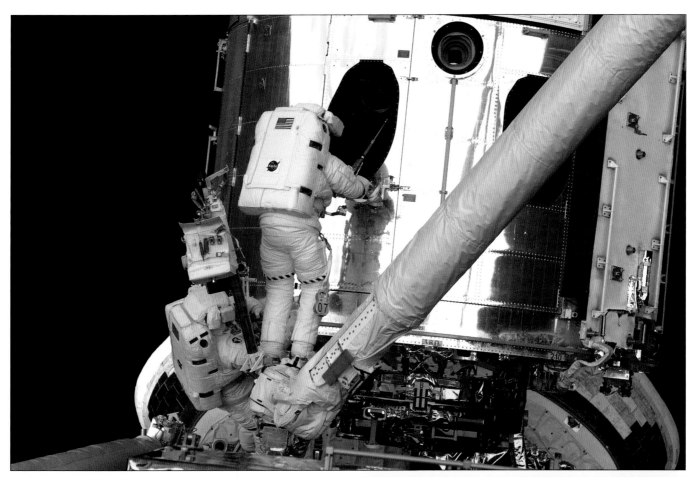

Michael Good, in the manipulator foot restraint at the end of the RMS arm, prepares to open a bay door on Hubble during EVA2 in order to swap out six gyros. The gyros are packaged in pairs called rate sensing units (RSU). Historically, Hubble's gyros have had a fairly high failure rate. By replacing all six, STS-125 has hopefully positioned Hubble for many more years of productive science. (NASA)

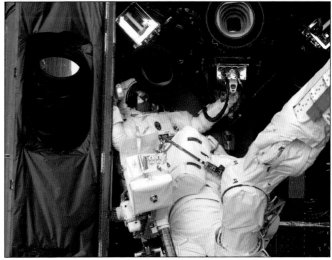

With the bay door open, Mike Massimino goes to work inside Hubble. Operating inside Hubble is always a tricky job, with tight quarters and delicate hardware. In particular for this EVA, note the star tracker light shade assemblies next to Massimino's helmet. The light shades are very fragile and Massimino was careful not to touch them. (NASA)

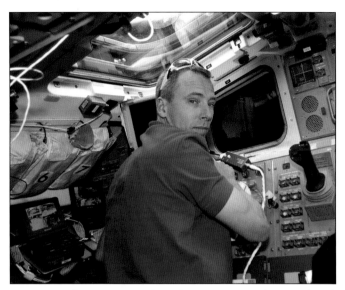

Drew Feustel supports EVA2 as the IVA crewmember on the aft flight deck. Due to crew exercise requirements, the IVA task was often shared between two crewmembers; John Grunsfeld started EVA2 while Feustel exercised, then the two swapped roles later in the EVA so Grunsfeld could exercise. (NASA)

Michael Good working on the rate sensing units from the outside of Hubble using the manipulator foot restraint as a work platform. Note the extensive use of foil insulation on the panels above where Good is working. (NASA)

Michael Good (in foot restraint) and Mike Massimino in the payload bay during EVA2. The starboard OMS pod is visible in the background, and in the lower left is the orbiter boom sensor system (OBSS) used to inspect the thermal protection system (TPS). (NASA)

Megan McArthur was the prime RMS arm operator on four of the five EVAs; Scott Altman relieved her for the other. The RMS controller is the smaller of the joysticks shown toward the left of the photo. Note the TV camera mounted near the elbow of the arm. (NASA)

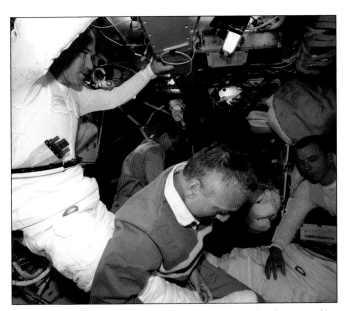

John Grunsfeld gets help from Mike Massimino in the donning of his Extravehicular Mobility Unit (EMU) spacesuit on the middeck in preparation for EVA3 on Flight Day 6. Drew Feustel is at right and Michael Good works in the background. (NASA)

Drew Feustel in his liquid cooling and ventilation garment (LCVG). The LCVG is a set of nylon tricot and spandex "long underwear" that is laced with thin plastic tubes. Cool water flows through these tubes to remove the heat produced by the astronaut. (NASA)

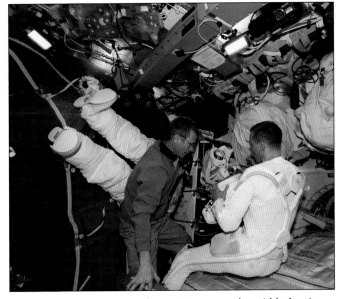

Greg Johnson assists Feustel in suiting up on the middeck prior to EVA3. The main cooling tubes are clearly visible on the back and legs; smaller tubes branch off to surround the body. The cooling water comes from the primary life-support system (PLSS) backpack. (NASA)

John Grunsfeld (left) and Drew Feustel prepare for EVA3 on 16 May. The EMU is a very sophisticated spacesuit manufactured by International Latex Corporation (ILC) in Dover, Delaware. The suit has 13 layers of material, including an inner cooling garment (two layers), pressure garment (two layers), thermal micrometeoroid garment (eight layers), and outer cover (one layer). All of the layers are sewn and cemented together to form the suit. In contrast to early space suits, which were individually tailored for each astronaut, the EMU has component pieces of varying sizes that can be put together to fit any given astronaut. Scott Altman is shown in the photo at right assisting his crewmates. (NASA)

John Grunsfeld, still wearing the pants from his EMU, poses in the commander's seat after EVA3. Note the circulation tube on his liquid cooling and ventilation garment and the red stripes on the legs of his spacesuit. (NASA)

Michael Good on the aft flight deck during FD6 activities. The yellow switches near the larger hand controller are for the floodlights in the payload bay. Note that each window is numbered with a small black label; 7 is the overhead and 9 looks into the payload bay. (NASA)

Servicing the Hubble Space Telescope

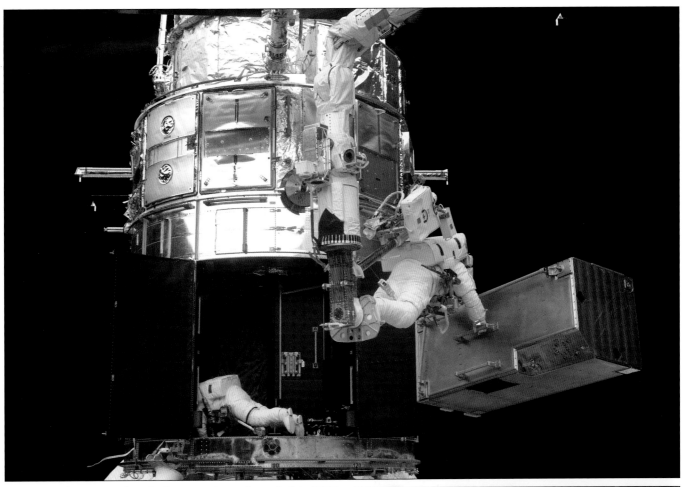

Drew Feustel, positioned on the foot restraint at the end of the RMS arm during EVA3, moves the Corrective Optics Space Telescope Axial Replacement (COSTAR). The doors to axial bay 4 are open and John Grunsfeld is partially inside. COSTAR had been installed by the crew of the first servicing mission on STS-61 to correct the flaw in the primary mirror. All of the replacement instruments have corrective optics built-in, so COSTAR was no longer needed and was replaced by the Cosmic Origins Spectrograph (COS). (NASA)

Drew Feustel (in foot restraint) and John Grunsfeld work with power tools in the payload bay of Atlantis during EVA3. Pretty much the definition of a cool job! (NASA)

Drew Feustel positioned on the manipulator foot restraint during EVA3. The empty axial bay 4, at left on Hubble, is ready to receive the Cosmic Origins Spectrograph (COS). The COS is an ultraviolet spectrograph optimized for observing faint extragalatic and galactic sources. Science goals include the study of the origins of large scale structure in the Universe, the formation and evolution of galaxies, and the origin of stellar and planetary systems and the cold interstellar medium. Both of the Orbiter's OMS pods are partially visible on either side of the photo. (NASA)

Servicing the Hubble Space Telescope

Mike Massimino peers through a window on the aft flight deck during EVA4 on Flight Day 7. John Grunsfeld, inside the Orbiter, appears amused by his crewmate. (NASA)

Michael Good during the 8-hour, 2-minute EVA4. Good and Mike Massimino (out of frame) worked on the Space Telescope Imaging Spectrograph (STIS). (NASA)

The Cape Canaveral Air Force Station and KSC, as seen from Atlantis on Flight Day 7. The Shuttle Landing Facility is located near the right top, with the VAB and two LC-39 launch pads below it. The Air Force Station is the point of land at the bottom. (NASA)

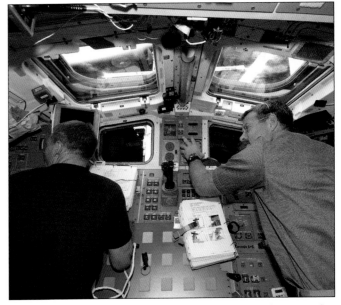

Scott Altman (right) operates the RMS for the repair of the STIS during EVA4 while Greg Johnson (left) performs photo/TV ops. It is not unusual for the commander to train as an arm operator on flights with extensive robotic arm operations. (NASA)

Michael Good, positioned on the RMS foot restraint and Mike Massimino (bottom) during EVA4. The Orbiters were originally manufactured with the capability of carrying two RMS arms, one on either payload bay sill. This capability as never used and only the port device was fitted. However, since the Columbia accident, the starboard sill has been used to carry the orbiter boom sensor system (OBSS) arm, which is generally similar to an RMS arm (and is shown at the left of the photo in its berthed position. (NASA)

John Grunsfeld and Megan McArthur review the payload deployment and retrieval system (PDRS) checklist during EVA4. (NASA)

Michael Good dons his spacesuit prior to EVA4. His right hand is resting on the crew escape pole that assists the crew to evacuate the Orbiter in the event of a bailout during low-speed gliding flight. When deployed, the pole extends through the crew hatch to ensure the astronauts do not recontact the vehicle after they bail out. (NASA)

To the commander goes the privilege of sleeping on the flight deck. Scott Altman operates a laptop PC while in a sleeping bag stretched across the commander and pilot seats at the end of Flight Day 7. Note the bag of dried fruit to the right of his head. (NASA)

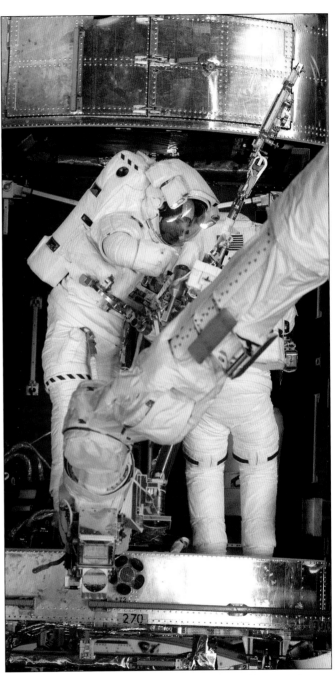

Michael Good (left, in foot restraint) and Mike Massimino (right, partially obscured by RMS) work inside Hubble during EVA4. The pair was working on the Space Telescope Imaging Spectrograph (STIS) located in radial bay 4. (NASA)

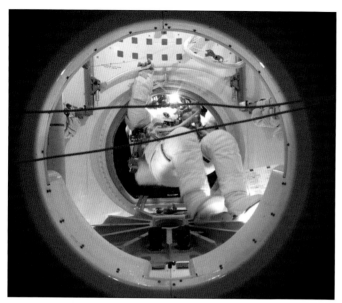

An unusual view of Michael Good egressing the external airlock at the start of EVA4. This photo was taken through the window on the hatch separating the airlock from the middeck. Note the open hatch inside the airlock at the bottom. (NASA)

John Grunsfeld (left) and Drew Feustel serve as IVA crewmembers on the flight deck of Atlantis during EVA4. The overhead control panels of the Orbiter, covered with hundreds of switches and circuit breakers, are visible at top. (NASA)

Megan McArthur prepares to change a lithium hydroxide (LiOH) canister on the middeck during Flight Day 9. These canisters absorb carbon dioxide and remove odors from the cabin air. As each canister is saturated it is replaced by a fresh canister. (NASA)

Greg Johnson floats through a hatch on the middeck during Flight Day 9. Note the floating EVA checklists and the fire extinguisher mounted to the bulkhead over the hatch. The light blue patches are pieces of velcro the crew can use to secure objects. (NASA)

Servicing the Hubble Space Telescope

A wide angle view of the forward flight deck of Atlantis taken during Flight Day 8. Atlantis was the first Orbiter in the fleet to receive the multifunction electronic display system (MEDS), or "glass cockpit" upgrade. Note that all but one of the MEDS displays are powered off to conserve the Orbiter's supply of cryogenic oxygen and hydrogen, used by the fuel cells to generate electrical power. (NASA)

The STS-125 crewmembers pose for a photo on the flight deck. On the front row (from left) are Scott Altman, Greg Johnson and Megan McArthur. Pictured on the back row (from left) are Michael Good, Mike Massimino, John Grunsfeld, and Drew Feustel. (NASA)

John Grunsfeld rides the manipulator foot restraint (MFR) on the end of the RMS during EVA5 on Flight Day 8. (NASA)

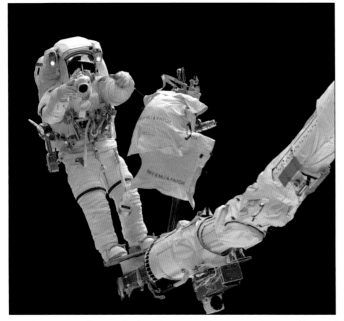

Another view of Grunsfeld in the MFR, this time taking a photo of Drew Feustel (who took this photo). EVA tool bags are secured to the MFR at right. (NASA)

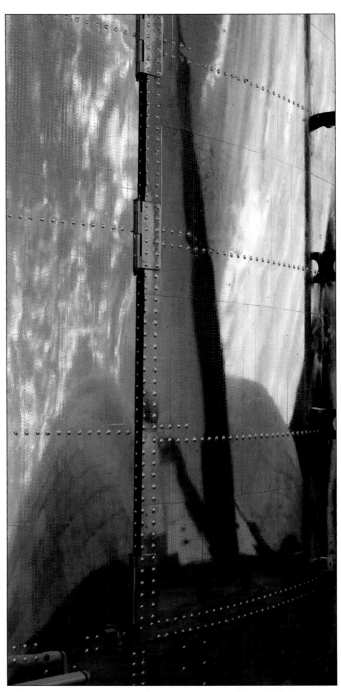

Atlantis' vertical stabilizer and OMS pods are reflected off the mirror-like surface of one of Hubble's bay doors. The black dots next to the vertical stabilizer are exhaust ports for an auxiliary power unit. (NASA)

The forward fuselage and payload bay of Atlantis, as photographed by a crewmember on the end of the RMS during EVA5. Visible at left is the RMS shoulder attached to the port payload bay sill. On the starboard sill are the orbiter's Ku-band antenna and the OBSS. (NASA)

Drew Feustel at work in the payload bay during EVA5. The RMS is visible at left and the bottom of Hubble at upper right. The silver object to the right of Feustel is the IMAX 3D camera that recorded the servicing mission for a future movie. (NASA)

The Orbiter Docking System (ODS) as photographed during EVA5. The docking mechanism was removed for STS-125 and replaced with a thermal blanket over the top of the airlock. The centerline docking camera can be seen through a circular hole in the blanket. (NASA)

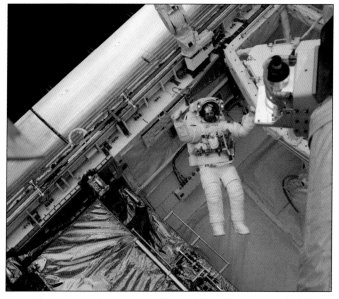

Drew Feustel participates in the fifth and final EVA. During the 7-hour and 2-minute spacewalk, Feustel and John Grunsfeld (out of frame) installed a battery module, replaced a fine guidance sensor and installed three NOBL thermal blankets. (NASA)

Mike Massimino (left) and Michael Good in their sleeping bags, which are attached to the lockers on the middeck. Note the Snoopy astronaut doll floating near Massimino's head. (NASA)

Drew Feustel working at the base of Hubble during EVA5. The left OMS pod of Atlantis is visible in the background. Note the strap connected from Feustel to the EVA safety line running along the payload bay sill. (NASA)

The servicing mission complete, Megan McArthur lifts Hubble out of the payload bay prior to release on Flight Day 9. Note the old-style NASA "worm" logo at the top of Hubble, one of the last surviving instances of a logo that Administrator Dan Goldin retired. (NASA)

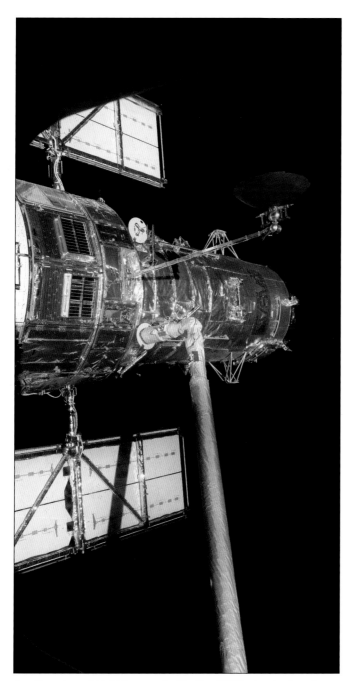

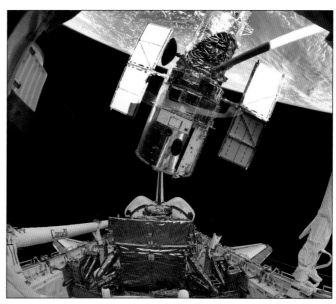

Once Hubble cleared the payload bay, Megan McArthur maneuvered the RMS for a test of the relative navigation system (RNS), which, along with the soft capture mechanism (SCM) is designed to facilitate a future robotic capture and deorbit of Hubble. (NASA)

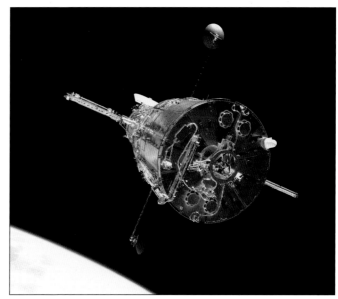

Another view of Hubble grappled by the RMS just prior to release. Note that the high-gain antennas have been deployed but the antenna dish is slewed to one side to provide clearance for the RMS during release. (NASA)

After McArthur released Hubble, Greg Johnson performed two reaction control system (RCS) separation burns designed to move Atlantis safely away from the telescope. This photo was taken from the aft flight deck between the first and second burns. (NASA)

A SLIGHT DETOUR

After releasing Hubble, STS-125 was still far from over. Two hours after release, Scott Altman and Greg Johnson performed an orbital adjust maneuver using the orbital maneuvering system (OMS) engines. This maneuver placed the Orbiter into an elliptical orbit, with the high point (apogee) at the altitude of Hubble, 350 miles, and the low point (perigee) at 184 miles. This maneuver reduced the risk of *Atlantis* being struck by micrometeoroids or orbital debris (MMOD). At the altitude of Hubble, the risk of MMOD is higher because lower atmospheric drag means that debris remains in orbit longer. The crew spent much of the remainder of the day performing a late inspection of the Orbiter thermal protection system (TPS) using the RMS and OBSS.

The following day was a much-deserved off-duty period for the crew. On the ground, KSC personnel started launch countdown procedures for STS-400 to allow *Endeavour* to be

Rain water saturates the ground and stands on the surface of a parking lot across the street from VAB. Thunderstorms in the vicinity of the Shuttle Landing Facility prevented Atlantis *from landing at KSC on Friday or Saturday, forcing the Orbiter to divert to Edwards AFB, California, on Sunday, 24 May.* (NASA/Jack Pfaller)

launched more quickly should the late inspection reveal damage requiring a rescue mission.

The crew returned to duty the following day with routine cabin stow procedures and a checkout of the flight control systems. After a thorough assessment of the data from the late inspection, NASA cleared *Atlantis* for entry. The STS-400 countdown was halted and its crew released from quarantine. The payloads for STS-127 were transported to LC-39A and preparations began for the rollaround of *Endeavour* from LC-39B.

STS-125 was supposed to land at the KSC Shuttle Landing Facility at 10:02 am on Friday, 22 May. But Mother Nature had other ideas. An unusual, severe thunderstorm system was pounding Central Florida with rain – over 25 inches during a three-day period in some areas. The Orbiter TPS, particularly the fragile white and black tiles, is not meant to fly through rain, and especially not through the frozen rain that is frequently present in thunderstorm clouds. Naturally, NASA had contingency plans, and this one was particularly simple: stay in space an extra day. However, when Saturday came, the weather in Florida had not changed. Officials debated sending *Atlantis* to its alternate landing site at Edwards Air Force Base (AFB) in California, but transporting the Orbiter back to Florida costs approximately $2 million and takes more than a week, so NASA elected to postpone landing another day.

Sunday, 24 May dawned with broken skies and scattered thunderstorms over Central Florida, and the forecast for Monday was not much better. NASA finally decided to land at Edwards, where the weather was perfect. At 9:25 am CDT, on its 197th orbit, while flying upside down and backward, Scott Altman fired the OMS engines for 2 minutes and 36 seconds to slow the Orbiter down for entry. About a minute into the burn, Mission Control and the crew noted a higher-than-normal pressure in a drain line on one of the auxiliary power units (APU) that provide hydraulic power to the aerosurfaces, rudder speedbrake, brakes, landing gear and nose wheel steering system. Engineers on the ground quickly determined the anomaly was not critical and that all systems were operating satisfactorily for landing. They were right; *Atlantis* had an uneventful entry and landing at Edwards.

STS-125 was the first Space Shuttle mission to end on the new concrete runway at Edwards. The Orbiter lands at approximately 195 knots and the brakes absorb some 36.5 million foot-pounds of energy during a normal stop. (Above: NASA/Jim Ross; Below: NASA/Tony Landis)

The drag chute was deployed at approximately 195 knots, and was jettisoned as the Orbiter slowed to 60 knots. A 9-foot-diameter pilot chute pulls the 40-foot-diameter main chute out of its compartment under the rudder. The chute trails the Orbiter by 89.5 feet. (NASA/Carla Thomas)

Atlantis comes home. Note the split rudder-speedbrake is partially opened as the Orbiter touches down. The Orbiter does not have any air-breathing engines and during entry and landing, the Orbiter is the world's heaviest and fastest glider – the pilots have to "get it right" since there is only one chance to land. (NASA/Tony Landis)

The crew on the runway: from left are Mike Massimino, Greg Johnson, Scott Altman, Megan McArthur, John Grunsfeld, Drew Feustel, and Michael Good. The crew wears their ACES pressure suits during entry, and changes into more comfortable flight suits before exiting the vehicle. (NASA/Tony Landis)

Atlantis *being towed to the NASA Dryden Flight Research Center. The large trailers connected to the Orbiter provide cooling and purge air to the APUs and fuel cells. In the bottom photo, the Orbiter is towed past the Boeing 747 shuttle carrier aircraft (SCA), N911NA, that will ferry it back to Florida.* (NASA/Jim Ross)

The convoy escorting Atlantis from the main Edwards runway to the NASA Dryden facility. The majority of the engineers and technicians that support landings at Dryden fly in from KSC, and only a minimal permanent staff is maintained. (NASA/Jim Ross)

Dryden has a Mate-Demate Device (MDD) that is very similar to the one at KSC. While in the MDD, the landing gear is retracted, systems are safed, some payload and crew items are removed, appropriate ballast (624 pounds in this case) is installed in the middeck, and the Orbiter is raised and mated to the SCA. (NASA/Tony Landis)

Atlantis departed Edwards AFB on 1 June on top of the second Shuttle Carrier Aircraft (SCA), N911NA. NASA operates two modified Boeing 747 jumbo jets as SCAs. The most obvious modifications are the small vertical endplates on the horizontal stabilizer and the mounting locations for the Orbiters. There are also numerous internal modifications. (Left: NASA/Tony Landis; Right: NASA/Jim Ross)

Following a refueling stop at Biggs Army Air Field (AAF), Texas, and an overnight stop at Columbus Air Force Base (AFB), Mississippi, Atlantis arrived at KSC at 6:53 pm on 2 June after making a low pass over the Indian River and the beach to allow the local public to see the mated pair. After touching down on the Shuttle Landing Facility, the pair was moved to the Mate-Demate Device (MDD). The photo at left shows technicians unsecuring the Orbiter from the SCA. (All: NASA/Jack Pfaller)

After it was moved to the MDD, Atlantis was unbolted from the SCA, the 747 moved away, and the Orbiter was then lowered to the ground on its landing gear. Note the SCA in the background of the photo at left. Late that night, Atlantis was towed to the high-bay in Orbiter Processing Facility One (OPF-1) to begin processing for STS-129, scheduled for 12 November. (All: NASA/Jack Pfaller)

Servicing the Hubble Space Telescope

EPILOG
CHANGING ASTRONOMY

The Hubble Space Telescope holds a unique position in history. It is one of the most popular scientific and engineering achievements of our era, and also one of the most important observatories ever built. Over 6,000 papers based on Hubble data have been published in peer-reviewed journals, and countless more have appeared in conference proceedings and the popular press. In the scientific world, one measure of an academic paper's success is how many times it is cited in subsequent articles: about one-third of all astronomy papers have no citations, while only two percent of papers based on Hubble data have no citations. On average, a paper based on Hubble data receives about twice as many citations as papers based on non-Hubble data.

Every achievement, however, comes at a cost. One study on the relative impacts on astronomy of different size telescopes found that while papers based on Hubble data generate 15 times as many citations as an equivalent ground-based instrument such as the William Herschel Telescope, Hubble cost about 100 times as much to build and maintain. The total cost of Hubble is difficult to estimate because of the NASA accounting system, but in gross terms it includes $2.5 billion for development, $500 million for launch, and another $2.5 billion for the five servicing missions. This does not include the cost of operating the STScI or TDRSS, the replacement science instruments, or the abortive robotic servicing missions.

The policies that govern the telescope have undoubtedly contributed to its incredible productivity. Hubble is an instrument for the entire astronomical community – any astronomer can request time on the telescope. Once observations are completed, the astronomers have a year to pursue their work before the data is released to the entire scientific community. Because everyone gets to see the data, the observations have given rise to a multitude of findings – many in areas that would not have been predicted by the original proposals.

Hubble has shown astronomers galaxies in all stages of evolution, including toddler galaxies that were around when the Universe was still young, helping scientists understand how galaxies form. It found protoplanetary disks, clumps of gas and dust around young stars that likely function as birthing grounds for new planets. Hubble has also discovered gamma-ray bursts – strange, incredibly powerful explosions of energy – that occur in far-distant galaxies when massive stars collapse.

Hubble has helped resolve some long-standing questions in astronomy, as well as turning up results that have required new theories to explain them. Among its primary mission was to measure distances to Cepheid variable stars more accurately than before, and thus constrain the value of the Hubble constant, the measure of the rate at which the Universe is expanding. Earlier estimates of the Hubble constant typically had errors of up to 50 percent, but measurements of Cepheid variables in the Virgo Cluster and other distant galaxy clusters provided an observed value with an accuracy of 10 percent. The Hubble constant also relates to the age of the Universe; a new estimate of 13 to 14 billion years is much more accurate than the previous range of 10 to 20 billion years.

Hubble has also been responsible for a wide range of observations that have cast doubt on existing theories and caused scientists to develop new theories. Perhaps the most significant of these is that instead of decelerating under the influence of gravity, the expansion of the Universe may in fact be accelerating. The high-resolution spectra and images provided by the Hubble have been especially well suited to establishing the prevalence of black holes in the nuclei of nearby galaxies. With the improvements made during STS-125, Hubble is poised to continue its leading-edge research for, perhaps, another decade. Whatever may happen, Hubble will undoubtedly contribute its usual phenomenal false-color images, further exciting the public imagination.

Earth, as seen from Atlantis during STS-125. (NASA)

The Final Servicing Mission